Microbiology: A Laboratory Experience

Microbiology: A Laboratory Experience

HOLLY AHERN

OPEN SUNY TEXTBOOKS

ISBN: 978-1-942341-54-3 ebook
978-1-942341-53-6 print

This publication was made possible by a SUNY Innovative Instruction Technology Grant (IITG). IITG is a competitive grants program open to SUNY faculty and support staff across all disciplines. IITG encourages development of innovations that meet the Power of SUNY's transformative vision.

Published by Open SUNY Textbooks, Milne Library
State University of New York at Geneseo,
Geneseo, NY 14454

Contents

Reviewer's Notes 1
Gail E. Rowe, Ph.D., Professor of Biology, La Roche College

Introduction 2

Main Body

Biosafety Practices and Procedures for the Microbiology Laboratory 5
The Microscopic World 10
Bacteriological Culture Methods 20
The Human Skin Microbiome Project 33
Differential Staining Techniques 49
Metabolism, Physiology, and Growth Characteristics of Cocci 58
Metabolism, Physiology, and Growth Characteristics of Bacilli 71
Microbiological Food Safety 79
The War on Germs 84
Epidemiology and Public Health 94
Blood: The Good, the Bad, and the Ugly 102
Microbe Mythbuster 116

Appendix A 121
Appendix B 126
Appendix C 128
Image Credits 130

Reviewer's Notes

GAIL E. ROWE, PH.D., PROFESSOR OF BIOLOGY, LA ROCHE COLLEGE

This laboratory manual was designed to accompany an undergraduate microbiology lecture course. The author states that it works equally well for a biology majors' course or for a course with a health sciences emphasis. I think it would be best used to support a microbiology course for nursing or health sciences, since the focus is on microbes of the human body and microbial diseases. For example the Human Skin Microbiome Project uses the human body as the environment, and food microbiology is addressed as food safety from pathogens. The introduction to common shapes and sizes of bacteria uses the natural microbial community of the human body. This teaching strategy shows microbes in a community, which is more natural then providing students with pure cultures, and may also stimulate student interest by making the source personal (their own mouths). The choice of the human body as a source of bacteria for these exercises further supports use of this lab manual for health sciences or nursing students.

A particular strength of this textbook is the presence of brief activities and questions to be completed by students that are imbedded within the background information. This strategy will likely draw students into more active learning, rather than passively listening to a lab lecture or just reading (or not reading) the manual prior to lab class.

Microscopy was presented with a good balance between use of the microscope vs. the physics behind its function. The author included sufficient theory for students to understand how a microscope works so they can make adjustments for optimal viewing, without covering unnecessary technical detail that might overwhelm the students. Similarly, the section on metabolism presents a very nice overview of the subject; including sufficient theory related to the lab topic without excessive detail.

The author effectively uses scaffolding to teach students how to create a dichotomous key for bacterial identification. By "walking" the students through the steps related to identifying traits (colony morphology, Gram reaction, other staining, biochemical tests), the manual systematically builds from simple questions and answers to a more complex and integrated problem of creating the dichotomous key. This strategy helps students sort out what they need to ask themselves at each step of building a dichotomous key, rather than just instructing them in the various identification methods without guidance on how to organize the tests in a systematic manner.

Introduction

Microbiology is a field of science devoted to the study of organisms that are too small to see, and therefore an engaging laboratory experience is often the key to capturing students' interest. It was with this in mind that the course behind this book was first conceived and developed. The goal was to provide my undergraduate microbiology students with an engaging and meaningful laboratory experience that nurtured a sense of discovery and encouraged greater interest in microbiology as a topic, a field of study, or a career.

This lab manual is suitable for use in a general microbiology or bacteriology course at a two-year or four-year college or university, and works equally well for a course with a Health Science emphasis. The approach, which has been field tested by hundreds of microbiology students over several years, builds skills while reinforcing core microbiology concepts introduced in lecture. The curriculum builds from the ground up. It begins with an introduction to biosafety practices and work with biological hazards, basic but essential microscopy skills, and aseptic technique and culture methods, and builds to include more advanced methods. The progression includes a semester-long investigation of a bacterial isolate and culminates with a practical evaluation of all microbiology skills learned in the course.

Key features and learning objectives of this laboratory manual include:

- Incorporation of the American Society for Microbiology Recommended Curriculum Guidelines for Undergraduate Microbiology.
- A design that permits flexibility in the sequencing of lab activities to correlate with lecture schedule.
- Student training in BSL 1 and 2 containment practices, in compliance with CDC Biosafety in Microbiological and Biomedical Laboratories, 5ed.
- Development of core and foundational laboratory skills required for working safely and effectively with bacterial cultures.
- Exposure to investigational methods used in public health and epidemiology.
- Student-centered investigational projects that encourages students to think and work like a scientist.
- Compatibility in content to most major microbiology textbooks.

Although the sequence in which the labs are done can be varied to suit individual course schedules, some are best ordered to follow a preceding lab. With this in mind, a schedule that works well for a standard semester with 14 weeks of actual class time is provided below.

Suggested Semester Lab Schedule:

Week 1: Biosafety Practices and Procedures for the Microbiology Laboratory

Week 2: The Microscopic World

Week 3: Bacteriological Culture Methods (begin primary culture for HSMP)

Week 4: The Human Skin Microbiome Project (including an introduction to Bergey's Manual and how to build and use a dichotomous key)

Week 5: Differential Staining Techniques

Week 6: Metabolism, Physiology, and Growth Characteristics of Cocci

Week 7: Lab Practical 1

Week 8: Metabolism, Physiology, and Growth Characteristics of Bacilli

Week 9: Microbiological Food Safety (includes selective/differential media and other physiological tests)

Week 10: Lab time for EI completion OR Microbe Mythbuster presentations

Week 11: Germ Warfare

Week 12: Epidemiology and Public Health

Week 13: Blood: The Good, the Bad, and the Ugly

Week 14: Lab Practical 2

The schedule above includes time for two lab practical exams to evaluate both acquired skills and understanding of concepts. Other assessment options include having students turn in their completed labs, or writing lab summaries (reports) as assignments each week once the lab activities are completed. For the Human Skin Microbiome Project (HSMP), the worksheets may be uploaded to a course management system for students to download and complete before handing in. Similarly, the activities included in other labs can be converted to documents that can be uploaded for student access, and used for assessment opportunities. Lab periods not allocated for lab practical can be used as library time for the Microbe Mythbuster project and student presentations.

Because so many of the techniques used are unique to a microbiology laboratory, demonstrating methods such as aseptic techniques, specific inoculation methods, and appropriate culture disposal methods to students before they perform them has proven to be a very effective way to instill confidence that they are performing the techniques correctly. This also has the additional benefit of improving lab safety and minimizing risks for students as they work with bacterial cultures.

For help with lab preparation and organization, appendices at the end of this book contain information on what supplies and equipment are needed for each lab, along with a list of suggested bacterial cultures. Since many established undergraduate microbiology labs may already have stock cultures of other organisms, the choice of which cultures to use for most labs is left up to the instructor. For some labs in which specific outcomes are necessary to illustrate a particular concept, cultures known to provide the desired outcome are specified. Cultures can be obtained through the American Type Culture Collection with ordering information available at http://www.atcc.org. Cultures are also available through several biological supply companies.

For the bacterial cultures suggested, expected outcomes for the microbiological tests students will be performing are given. These outcomes are based on the information provided in *Bergey's Manual of Determinative Bacteriology*, 9th edition, which was originally developed to facilitate identification of bacterial species. Since 1992, *Bergey's Manual of Systematic Bacteriology*, which has an expanded scope and encompasses relatedness among species (systematics) in addition to identification, has largely replaced the older text and has become the definitive reference on bacterial taxonomy.

Bergey's Manual of Determinative Bacteriology (*"Bergey's Manual"*) is the primary reference for the phenotypic tests and morphological characteristics of the cultures suggested for use in this book, and the basis for the Human Skin Microbiome Project, in which a bacterial isolate is presumptively identified to genus and species using traditional bacteriological culture methods and phenotypic tests. Although automated and DNA-based tests are replacing this approach in clinical laboratories, for this lab experience the emphasis is on the investigative process and *Bergey's Manual* fits the need perfectly. The book is available in paperback form (from Amazon and Carolina), and it is suggested that it be made available as a resource students can use while in the laboratory.

One final consideration is the lab time. Although all of the labs can be completed in three- hour lab periods, many require incubation time to permit bacterial cultures to metabolize, reproduce, and grow colonies or other forms of culture growth. To accommodate this need, we have made available additional time for students to return to lab and complete experiments. "Open Lab" times are supervised by laboratory assistants (most are work-study students who successfully complete the course) who have been trained to provide technical assistance to the course by making culture media, prepping and cleaning up after the lab each week, and serving as peer tutors. Although most students have been able to take advantage of the Open Lab time, for campuses with a large population of students who commute or have families and/or jobs, coming back to lab the next day isn't always a possibility. We have solved this by giving students the option of incubating cultures at a lower temperature if they know they will be unable to return to lab; students can also arrange to have their cultures transferred to a refrigerator to preserve the results of their tests.

Biosafety Practices and Procedures for the Microbiology Laboratory

The following recommended practices and procedures for working safely on microbiology projects in a teaching laboratory environment are based on "*Guidelines for Biosafety in Teaching Laboratories*," from the American Society for Microbiology (ASM). The full documents may be viewed at:

http://www.asm.org/images/asm_biosafety_guidelines-FINAL.pdf

http://www.asm.org/images/Education/FINAL_Biosafety_Guidelines_Appendix_Only.pdf

What is Biosafety?

Microbiology is a science that investigates the biology of microscopic organisms. Although individual cells of these organisms may be directly observed with a microscope, and their shapes and activities observed, to investigate other characteristics such as metabolism or genetics, growing cells in populations (called **cultures**) is the preferred approach. For many types of experiments, all of the cells in the population must be essentially the same; such populations are called **pure cultures**. A set of techniques, mostly developed in the late 19th century by Robert Koch, Louis Pasteur, and their collaborators, permitted the isolation of bacteria from their natural environments and separation into pure cultures for further study. Probably more than anything else, these are the techniques that define microbiology as a scientific field of study.

It has been estimated that less than 1% of bacteria can be grown in culture in a laboratory. However, many types can be, and those are the ones on which we will focus. Microbiologists culture bacteria by providing them with food, water, and other growth requirements in an environment with a constant and comfortable growth temperature. These requirements vary, depending on the natural growth conditions for the microbial populations under study.

Food is provided in the **culture media** we use. Media may be in the form of a liquid (called a "broth") or solid or semi-solid forms, either in tubes or in culture dishes (Petri plates). The choice of media depends on what you want to do or need to know about the bacteria in your cultures.

To ensure that we culture only the specific bacteria we want, and nothing else from the environment, microbiologists use a set of strict **aseptic techniques**, which protects us from the bacteria in the cultures, and also protects our cultures from contaminants in the environment. These methods will be demonstrated in class. In addition, specific laboratory rules must be followed for containment of microbial cultures in the laboratory, for the safety of all. For this laboratory, these practices are listed below.

Laboratory Safety Practices and Procedures

1. Remember that all bacteria are potential pathogens that may cause harm under unexpected or unusual circumstances. If you as a student have a compromised immune system or a recent extended illness, you should share those personal circumstances with your lab instructor.
2. Know where specific safety equipment is located in the laboratory, such as the fire extinguisher and the eyewash station.
3. Recognize the international symbol for biohazards, and know where and how to dispose of all waste materials, particularly biohazard waste. Note that all biohazard waste must be sterilized by autoclave beforeit can be included in the waste stream.

4. Keep everything other than the cultures and tools you need OFF the lab bench.
5. All of the equipment and supplies used in experiments involving bacterial cultures should be sterilized. This includes the media you use and also the tools used for transferring media or bacteria, such as the inoculating instruments (loops and needles) and pipettes for liquid transfer.
6. Transfer of liquid cultures by pipette should NEVER involve suction provided by your mouth.
7. Disinfect your work area both BEFORE and AFTER working with bacterial cultures.
8. In the event of an accidental spill involving a bacterial culture, completely saturate the spill area with disinfectant, then cover with paper towels and allow the spill to sit for 10 minutes. Then carefully remove the saturated paper towels, dispose of them in the biohazard waste, and clean the area again with disinfectant.
9. Wear gloves when working with cultures, and when your work is completed, dispose of the gloves in the biohazard garbage. Safety glasses or goggles are also recommended.
10. Long hair should be pulled back to keep it away from bacterial cultures and open flame.
11. Make sure that lab benches are completely cleared (everything either thrown away or returned to storage area) before you leave the lab.

Bacteria pose varying degrees of risk both in a controlled laboratory environment and in their natural settings. Therefore, the level of containment necessary for working safely with bacterial cultures also varies according to a system that classifies microbes into one of four biosafety levels (BSL), which provides minimum standards for safe handling of microbes at each level. BSLs are defined and containment practices are detailed by the Centers for Disease Control and Prevention (CDC) for laboratories in the United States. The full document, "*Biosafety in Microbiological and Biomedical Laboratories,*" can be viewed in its entirety at http://www.cdc.gov/biosafety/publications/bmbl5/index.htm.

Prior to the first meeting of the laboratory, navigate to the following web page to review the BSL guidelines as outlined by the CDC, which includes an interactive quiz:

http://www.cdc.gov/training/Quicklearns/biosafety/

Most of the bacterial cultures we will be working with are classified as BSL-1. Below, list three practices that should be used while you are working with BSL-1 bacteria:

1. ______________________________

2. ______________________________

3. ______________________________

In a few of the labs, we will be working with bacteria that are classified as BSL-2. What additional safety practices should you employ when you work with these bacteria?

1. ______________________________

2. ______________________________

3. ______________________________

After your instructor has discussed any additional safety procedures for your lab, complete the biosafety concept assessment on the following page (or one provided by your instructor) and sign the affirmation. Return the completed document to your instructor.

Microbiology Laboratory Biosafety Guidelines

Complete the following questions by writing the letter of the correct answer on the line.

___ 1. Biological agents are assigned to Biosafety Levels (BSLs) based on

a. whether they are bacteria, viruses, or other microorganisms.

b. their susceptibility to laboratory disinfectants.

c. the risk they pose to human health and the environment.

d. the amount of the agent that will be used in the lab.

___ 2. BSL 1 practices, equipment, and facilities are used when working with which type of microorganism?

a. Bacteria that are well characterized and not known to consistently cause disease in healthy adult humans.

b. Bacteria that are considered moderate-risk agents present in the community and associated with human diseases.

c. Indigenous or exotic bacteria with potential for aerosol transmission and associated with human disease.

d. Dangerous/exotic agents with a high risk of life-threatening disease.

___ 3. The primary hazards to people working in a BSL 1 or 2 laboratory relate to

a. accidental skin punctures caused by mishandling of sharp lab tools.

b. mucous membrane exposure due to accidental splashing of cultures.

c. accidental ingestion of infectious material.

d. All of the above are primary hazards in a BSL 1 or 2 lab.

___ 4. Which of the following is required when working with BSL 2 agents?

a. Biohazard warning signage on doors and lab areas where agents are kept or used.

b. Self-closing, double door access to the lab facility.

c. Clothing change when entering or leaving the lab.

d. There are no special precautions needed for a BSL 2 lab.

___ 5. What is the appropriate disposal method for a bacterial culture grown on an agar plate?

a. Throw the culture away in any one of the several garbage containers in the room.

b. Throw the culture away in the garbage container marked with a biohazard sign.

c. Allow the agar to dehydrate in the incubator and then discard in any garbage.

d. Bacteria grown on an agar plate medium should never be disposed of.

Microbiology Laboratory Biosafety Affirmation

By signing your name below, you affirm that:

- Potential hazards in the microbiology laboratory have been identified and explained.
- You understand that the bacterial cultures used in lab require either BSL 1 or BSL 2 containment procedures.
- Safety practices for BSL 1 and 2 have been explained and/or demonstrated to you.
- You fully understand the hazards associated with BSL 1 and 2 bacterial cultures and are willing to assume the risk.
- You are responsible for your own actions while working in a laboratory with potentially hazardous materials.

PRINT your FULL NAME ____

SIGN your name ________________________________ Date ____________

The Microscopic World

The Compound Microscope

There would be little to do in a microbiology laboratory without a microscope, because the objects of our attention (bacteria, fungi, and other single celled creatures) are otherwise too small to see. Microscopes are optical instruments that permit us to view the microbial world. Lenses produce the magnified images that allow us to visualize the form and structure of these tiniest of living beings.

To use this important piece of equipment properly, it is helpful to know how a microscope works. A good place to begin is to learn the name and function of all of the various parts, because when we talk about the ways to improve microscopic images, terms like "ocular lenses" and "condenser" always come up.

Based on the picture of the binocular, compound light microscope in Figure 1, match the name of the major part (listed below) with its location on the microscope, and give a very brief description of what each is used for:

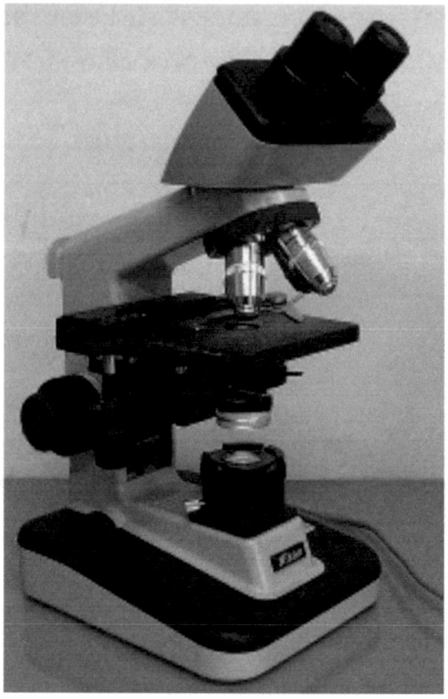

Figure 1

Ocular lenses ____________________

Objective lenses ____________________

(Revolving) Nosepiece ____________________

Stage and stage clips ____________________

Course and fine focus knobs ____________________

Condenser lens ____________________

Iris diaphragm ____________________

Locate the parts on the microscope that allow you to:

- Move the stage (stage adjustment knobs)
- Adjust the condenser lens
- Adjust the light intensity
- Adjust the iris diaphragm
- Adjust the distance between the ocular lenses

Making Images

With a bright-field microscope, images are formed as a result of the interplay between light waves, the object, and lenses. How images of biological objects are formed is actually more physics than biology. Since this isn't a physics course, it's more important to know how to create exceptional images of the object than it is to know precisely how those images are formed.

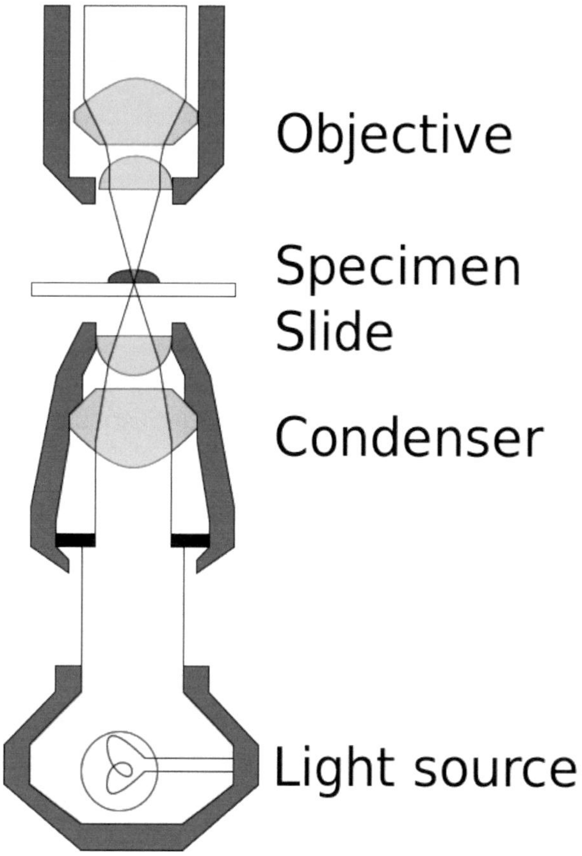

Figure 2

Light waves that pass through and interact with the object may speed up, slow down, or change direction as they travel through "media" (such as air, water, oil, cytoplasm, etc.) of different densities. For example, light passing through a thicker or denser part of a specimen (such as the nucleus of a cell) may be reflected or refracted ("bend" by changing speed or direction) more than those waves passing through a thinner part. This makes the thicker part appear darker in the image, while the thinner parts are lighter.

For a compound microscope, the optical path leading to a detectable image involves two lenses – the objective lens and the ocular lens. The objective lens magnifies the object and creates a **real image**, which will appear to be 4, 10, 40, or 100 times larger than the object actually is, depending on the lens used. The ocular lens further magnifies the real image by an additional factor of 10, to produce a vastly larger **virtual image** of the object when viewed by you.

Light from an illuminator (light source) below the stage is focused on the object by the condenser lens, which is located just below the stage and adjustable with the condenser adjustment knob. The condenser focuses light through the specimen to match the aperture of the objective lens above, as illustrated in Figure 2.

Appropriate use of the condenser, which on most microscopes includes an iris diaphragm, is essential in the quest for a perfect image. Raising the condenser to a position just below the stage creates a spotlight effect on the specimen, which is critical when higher magnification lenses with small apertures are in use. On the other hand, the condenser should be lowered when using the scanning and low power lenses because the apertures are much larger, and too much light can be blinding. For creating the best possible contrast in the image, the iris diaphragm can be opened to make the image brighter or closed to dim the light. These adjustments are subjective and should suit the preferences of the person viewing the image.

When the light waves that have interacted with the specimen are collected by the lenses and eventually get to your eye, the information is processed into dark and light and color, and the object becomes an image that you can see and think more about.

Magnification

The microscope you'll be using in lab has a compound system of lenses. The objective lens magnifies the object "X" number of times to create the real image, which is then magnified by the ocular lens an additional 10X in the virtual image. Therefore, the total magnification, or how much bigger the object will actually appear to you when you view it, can be determined by multiplying the magnification of the objective lens by 10.

The magnifying power of each lens is engraved on its surface, followed by an "X." In the table below, find the magnification, and then calculate the total magnification for each of the four lenses on your microscope.

	Magnification of objective lens	Total magnification of viewed object
Scanning Lens		
Low Power Lens		
High Power Lens		
Oil Immersion Lens		

Let's say you wanted to look at cells of *Bacillus cereus*, which are rod-shaped cells that are about 4 µm long. If you were observing *B. cereus* with a microscope using the high power lens, how big would the cells appear to be when you look at them? ______________________

Resolution Limits Magnification

So the microscope makes small cells look big. But why can't we just use more or different lenses with greater magnifying power until the images we see are really, really big and easier to see?

The answer is **resolution.** Consider what happens when you try to magnify the fine print from a book with a magnifying glass. As you move the lens away from the print, it gets larger, right? But as you keep moving the lens, you notice that while the letters are still getting larger, they are becoming blurry and hard to read. This is referred to as "empty magnification" because the image is larger, but not clear enough to read. Empty magnification occurs when you exceed the **resolving power** of the lens.

Resolution is often thought of as how clearly the details in the image can be seen. By definition, resolution is the minimum distance between objects needed to be able to see them as two separate entities. It can also be thought of as the size of the smallest object that we can clearly see.

The ability of a lens to resolve detail is ultimately limited by diffraction of light waves, and therefore, the practical limit of resolution for most microscopes is about 0.2 μm. Therefore, it would not be practical to try to observe objects smaller than 0.2 μm with a standard optical microscope. In addition, cells of all types of organisms lack contrast because many cellular components refract light to a similar extent. This is especially true of bacteria. To overcome this problem and increase contrast, biological specimens may be stained with selective dyes.

The Oil Immersion Objective

The lens with highest magnifying power is the oil immersion lens, which achieves a total magnification of 1000X with a resolution of 0.2 μm. This lens deserves special attention, because without it our time in lab would be frustrating.

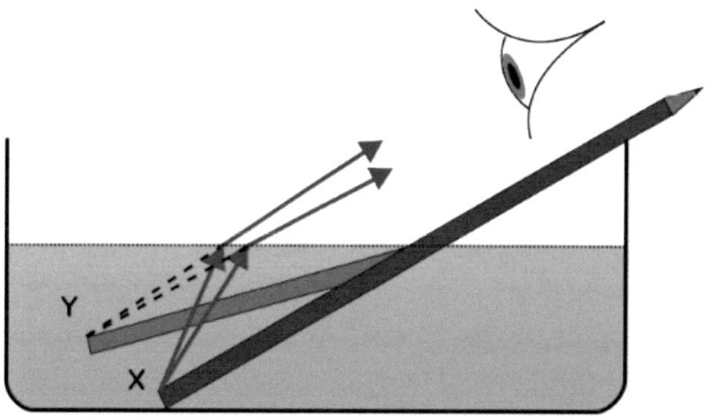

Figure 3

The resolving power of this lens is dependent on "immersing" it in a drop of oil, which prevents the loss of at least some of the image-forming light waves because of refraction. Refraction is a change in the direction of light waves due to an increase or decrease in the wave velocity, which typically occurs at the intersection between substances through which the light waves pass. This is a phenomenon you can see when you put a pencil in a glass of water. The pencil appears to "bend" at an angle where the air and water meet (see Figure 3). These two substances have different refractive indices, which means that light passing through the air reaches your eye before the light passing through the water. This makes the pencil appear "broken."

The same thing happens as the light passes through the glass slide into the air space between the slide and the lens. The light will be refracted away from the lens aperture. To remedy this, we add a drop of oil to the slide and slip the oil immersion objective into it. Oil and glass have a similar refractive index, and therefore the light bends to a lesser degree and most of it enters the lens aperture to form the image.

It is important to remember that **you must use a drop of oil** whenever you use the **oil immersion objective** or you will not achieve maximum resolution with that lens. However, **you should never use oil with any of the other objectives**, and you should be diligent about wiping off the oil and cleaning all of your lenses each time you use your microscope, because the oil will damage the lenses and gum up other parts of the instrument if it is left in place.

Using the Microscope

If you are new to microscopy, you may initially feel challenged as you try to achieve high quality images of your specimens, particularly in the category of "Which lens should I use?" A simple rule is: the smaller the specimen, the higher the magnification. The smallest creatures we observe are bacteria, for which the average size is a few micrometers (μm). Other microscopic organisms such as fungi, algae, and protozoa are larger, and you may only need to use the high power objective to get a good view of these cells; in fact, using the oil immersion objective may provide you with less information because you will only be seeing a part of a cell.

This brings us to two additional concepts related to microscopy—working distance and parfocality. **Working distance** is how much space exists between the objective lens and the specimen on the slide. As you increase the magnification by changing to a higher power lens, the working distance decreases and you will see a much smaller slice of the specimen. Also, once you've focused on an object, you should not have to make any major adjustments when you switch lenses, because the lenses on your microscope are designed to be **parfocal**. This means that something you saw in focus with the low power objective should be nearly in focus when you switch to a high power objective, or vice versa. Thus, for viewing any object and regardless of what lens you will ultimately use to view it, the best practice is to first set the working distance with a lower power lens and adjust it to good focus using the coarse focus knob. From that point on, when you switch objectives, only a small amount of adjustment with the fine focus knob should be necessary.

Here is a final consideration related to objective lenses and magnification. Look at the lenses on your microscope, and note that as the magnification increases, the length of the lens increases and the lens aperture decreases in size. As a result, you will need to adjust your illumination to compensate for a darkening image. There are essentially three ways to vary the brightness; by increasing or decreasing the light intensity (using the on/off knob), by moving the **condenser lens** closer to or farther from the object using the condenser adjustment knob, and/or by opening/closing the **iris diaphragm**. Don't be afraid to experiment to create the best image possible.

Guidelines for safe and effective use of a microscope:

1. Carry the microscope to your lab table using two hands, and set it down gently on the bench. Once placed on the bench, do not try to slide it around on its base, because this is extremely jarring to the optical system.

2. Clean all of the lenses with either lens paper or Kimwipes (NOT paper towels or nose tissues) **BEFORE** you use your microscope, **AFTER** you are done, and before you put it away.

3. When you are finished with the microscope, check the stage to make sure that you don't leave a slide clipped in the stage. Make sure to switch the microscope OFF before you unplug it. Gently wrap the cord around the base and cover your microscope with its plastic cover.

4. Return the microscope to the cabinet before you leave the lab. Make sure that the ocular lenses are facing IN.

Method

Together we will review how to effectively achieve an exceptional image using a standard optical microscope. This will include not only locating and focusing on the object, but also using the condenser lens and iris diaphragm to achieve a high degree of contrast and clarity.

Rectal Smear

We'll start by looking at a prepared slide of a "rectal smear," which is quite literally a smear of feces on a slide stained with a common method called the Gram stain. You will observe several different types of bacterial cells in this smear that will appear either pink or purple. While the main purpose of this is to develop proficiency in use of the oil immersion objective lens, it also provides the opportunity to look at bacteria, observe the differences in cell shapes and sizes, and note that when Gram stained they turn out to be either purple or pink.

When you have achieved an exceptional image of the fecal bacteria at 1000X, consider the following questions.

- In a single field, approximately what proportion of the bacterial cells are circular (the microbiology term for circular bacterial cells is "cocci")? ______________________

- Among the cells that are cocci, can you see any specific types of arrangements, like chains of cocci (called "streptococci") or clusters (called "staphylococci")? Sketch examples in the space below:

- In a single field, approximately what proportion of the bacterial cells are rod-like (the microbiology term for rod-shaped cells is "bacilli")? ______________________

- Among the cells that are bacilli, can you see any specific types of arrangements, like pairs (diplobacilli), chains (streptobacilli) or parallel clusters (palisades)? Sketch examples in the space below:

- Based on the shape/arrangement of the bacterial cells, and now including color (whether they are pink or purple), estimate how many different types of bacteria you are able to see in a single microscopic field. ______________________

What's in YOUR Mouth?

The human mouth is home to numerous microbes, which persist no matter how many times you brush your teeth and use mouthwash. Since these microbes generally inhabit the surface layers of the oral mucosa, we humans have evolved ways to keep their numbers under control, including producing antibacterial chemicals in saliva and constantly turning over the outer layer of epithelial cells that line the inside of the mouth.

Obtain a prepared slide labeled "mouth smear." On this slide you will see large cells with a nucleus, clearly visible with both the low power and high power objective lenses. These are squamous epithelial cells that form the outermost layer of the oral mucosa. At high power, you should start to see small cells on the surface of the larger epithelial cells. With the oil immersion objective lens, you will be able to tell the smaller cells are bacteria.

Locate and focus on a single squamous epithelial cell with obvious bacteria on its surface. Create a sketch of the "cheek" cell (as squamous epithelial cells are sometimes called) in the circle provided. Then label the cell membrane, cytoplasm, and nucleus of the "cheek" cell, which should be easily observed.

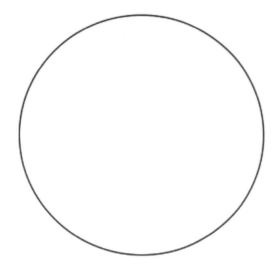

Add to your illustration the bacterial cells which you should see on or near the larger larger cheek cells. Try to keep the size of the bacterial cells to scale with the size of the cheek cell.

- How would you describe the shape and arrangement of the bacterial cells (using the microbiology terms you used to describe the bacteria in the rectal smear)?

- The nucleus of a squamous epithelial cells is approximately 10 µm in size. Compare the size of the cell's nucleus with the size of the bacterial cells. Based on this comparison, what is the approximate size of the bacterial cells in the image?

Once you've looked at the prepared slide, obtain a glass slide and a sterile swab. Collect a sample of your oral mucosa by gently rubbing the swab over the inside of your cheek. Smear the swab over the surface of the slide (this is known as making a "smear" in microbiology). Allow the smear to dry, and then heat fix by passing the slide through the flame of a Bunsen burner, as demonstrated. Discard the swab in the biohazard waste.

Once the sample is heat fixed, stain it with **safranin**. This is a pinkish-red colored stain, and all cells (both bacterial and your mouth cells) will take up the stain and increase the contrast in the image.

Observe your mouth smear with the microscope. When you get to the oil immersion objective, locate and focus on a single cheek cell. As you did with the prepared slide, sketch the larger cheek cell in the circle provided and label the membrane and nucleus . Add the bacterial cells to your sketch, and try to keep the size scale accurate.

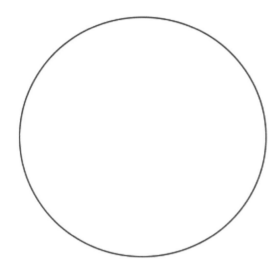

- Below, describe the shape and arrangement of the different types of bacterial cells you observe in the smear.

Will Yogurt Improve your Health?

Yogurt is produced when lactic acid (homolactic) bacteria that naturally occur in milk ferment the milk sugar lactose and turn it into lactic acid. The lactic acid accumulates and causes the milk proteins to denature ("curdle") and the liquid milk becomes viscous and semi-solid.

Within the past few years, positive health benefits have been correlated with eating fermented foods containing "live" cultures. Although several types of bacteria are known to ferment milk and produce yogurt, two genera in particular, *Lactobacillus* and *Bifidobacterium*, have been singled out as promoting good digestive health and a well-balanced immune response. Both of these are bacilli arranged in pairs or short chains. *Streptococcus* spp., which are cocci arranged in chains, are also usually involved in the process of making the milk into yogurt, but these are not directly associated with positive health benefits to the person who eats the yogurt.

Obtain a prepared slide labeled "yogurt smear" and view it with the microscope. The milk proteins in the yogurt will be visible as lightly stained amorphous blobs. By now you should have a pretty good idea of what bacteria look like, so locate and focus on areas where you see bacterial cells.

- Using microbiology terms, describe the shape and arrangement of the different types of bacterial cells you see in the smear

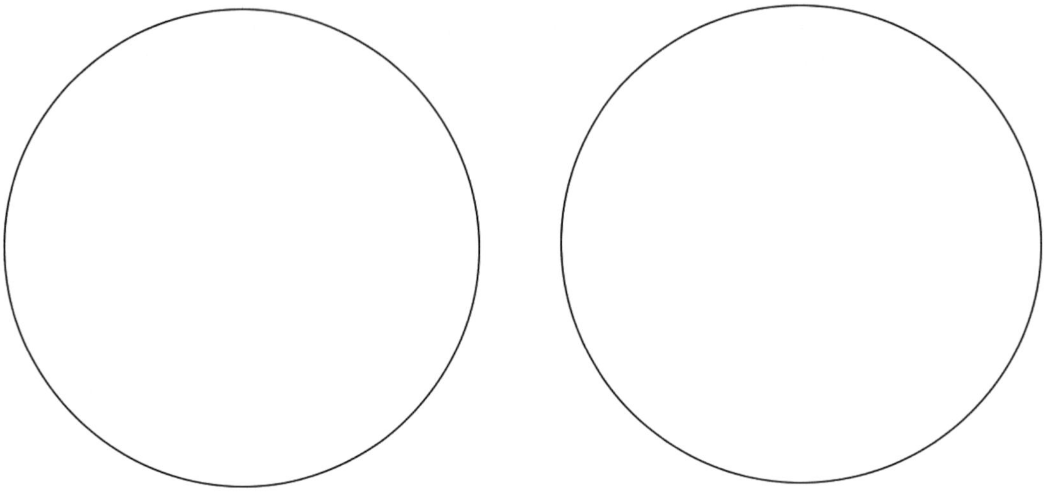

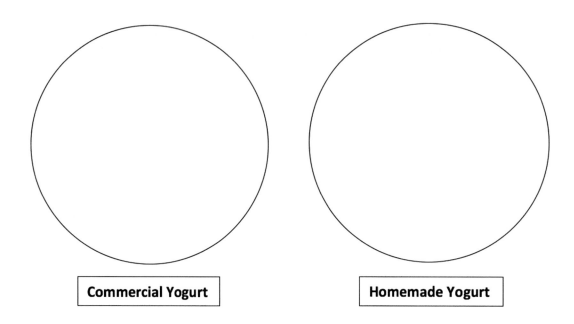

Commercial Yogurt **Homemade Yogurt**

Once you've made the observations using the prepared slide, obtain a glass slide and a sterile swab. Collect a sample from the container of commercially prepared yogurt by swirling the swab in the yogurt, then scraping of the excess on the edge of the container. Smear this over the surface of the slide, making sure that you leave only a thin film of yogurt on the surface. Make a second smear from the container of freshly prepared homemade yogurt, if available. Allow both smears to air dry, and then heat fix them.

Once the sample(s) are heat fixed, stain them with **crystal violet.** This is a purple colored stain, and although both the milk proteins and cells will stain this color, the milk stains faintly and the bacteria will appear dark purple. Keep in mind that the probiotic bacteria are bacilli. Below, sketch a representative field as seen with the oil immersion objective for each of the yogurt samples.

Move your stage so you observe 10 different microscope fields. Keep track of the number of different types of bacterial cells you encounter during your survey, and record that information below:

- Cell Count of Bacteria in Commercial Yogurt:
    - Number of cocci in 10 microscope fields: ___________
    - Number of bacilli in 10 microscope fields: ___________
- Cell Count of Bacteria in Homemade Yogurt:
    - Number of cocci in 10 microscope fields: ___________
    - Number of bacilli in 10 microscope fields: ___________
- Does the container of commercially prepared yogurt state that there are "live, active cultures" in the yogurt? If the container lists the name(s) of the bacteria, write them below, followed by whether they should be bacilli or cocci.

- Considering the relative number and type(s) of bacteria you saw in the stained smear of the commercially prepared yogurt, what do you conclude about the health benefits of eating products such as this as a probiotic?

- Considering the relative number and type(s) of bacteria you saw in the homemade yogurt, what do you conclude about the health benefits of eating a fresh, homemade type of yogurt as a probiotic?

Bacteriological Culture Methods

The Birth of Bacteriology

While perhaps best known to us as a cause of human disease, bacteria really should be far more famous for their positive contributions than for their negative ones. Below, list three positive things that bacteria do for you.

1. ______________________________
2. ______________________________
3. ______________________________

Bacteria were first observed by Anton von Leeuwenhoek in the late 17th century, but didn't become the objects of serious scientific study until the 19th century, when it became apparent that some species caused human diseases. The methods devised by Robert Koch, Louis Pasteur, and their associates during the "Golden Age" of microbiology, which spanned from the mid-1800s to early 1900s, are still widely used today. Most of these methods involved isolating single bacteria derived from a natural source (such as a diseased animal or human) and cultivating them in an artificial environment as a pure culture to facilitate additional studies.

During the middle of the twentieth century, when we believed we had defeated them at their disease-causing game, bacteria became popular subjects of empirical study in fields such as genetics, genetic engineering, and biochemistry. With the evolution of antibiotic-resistant strains and our increased knowledge of bacterial stealth attack strategies such as biofilms and intracellular growth, medical researchers have refocused their attention on disease-causing bacteria and are looking for new ways to defeat them.

Growing bacteria in pure culture is still one of the most widely used methods in microbiology. Many bacteria, particularly those that cause diseases and those used in scientific studies, are heterotrophic, which means that they rely on organic compounds as food, to provide energy and carbon. Some bacteria also require added nutritional components such as vitamins in their diet. An appropriate physical environment must be created, where important factors such as temperature, pH, and the concentration of atmospheric gases (particularly oxygen) are controlled and maintained.

The nutritional needs of bacteria can be met through specialized microbiological media that typically contain extracts of proteins (as a source of carbon and nitrogen), inorganic salts such as potassium phosphate or sodium sulfate, and in some cases, carbohydrates such as glucose or lactose. For **fastidious** bacteria (meaning, those that are picky eaters) vitamins and/or other growth factors must be added as well.

Bacteriological culture media can be prepared as a liquid (broth), a solid (plate media or slant media), or as a semi-solid (deeps) as illustrated in Figure 1. Solid and semi-solid media contain a solidifying agent such as agar or gelatin. Agar, which is a polysaccharide derived from red seaweed (*Rhodophyceae*) is preferred because it is an inert, non-nutritive substance. The agar provides a solid growth surface for the bacteria, upon which bacteria reproduce until the distinctive lumps of cells that we call colonies form.

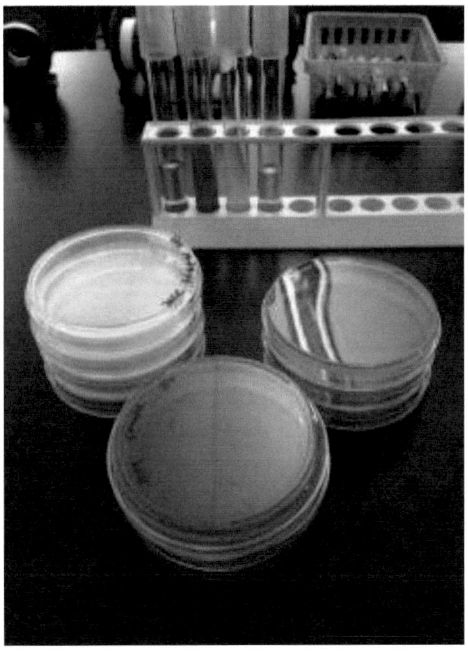

Figure 1. Different types of culture

Koch, Pasteur, and their colleagues in the 19th and early 20th centuries created media formulations that contained cow brains, potatoes, hay, and all sorts of other enticing microbial edibles. Today, bacteriological media formulations can be purchased in powdered form, so that all the preparer has to do is to measure out the correct amount, add the right amount of water, and mix. After the basic formula has been prepared, the medium is sterilized in an autoclave, which produces steam under pressure and achieves temperatures above boiling. Once sterilized media has cooled, it is ready to be used.

Growing Bacteria in Culture

A population of bacteria grown in the laboratory is referred to as a **culture.** A **pure culture** contains only one single type; a **mixed culture** contains two or more different bacteria. If a bacterial culture is left in the same media for too long, the cells use up the available nutrients, excrete toxic metabolites, and eventually the entire population will die. Thus bacterial cultures must be periodically transferred, or **subcultured**, to new media to keep the bacterial population growing.

Microbiologists use subculturing techniques to grow and maintain bacterial cultures, to examine cultures for purity or morphology, or to determine the number of viable organisms. In clinical laboratories, subculturing is used to obtain a pure culture of an infectious agent, and also for studies leading to the identification of the pathogen. Because bacteria can live almost anywhere, subculturing steps must be performed **aseptically**, to ensure that unwanted bacterial or fungal contamination is kept out of an important culture.

In microbiology, aseptic techniques essentially require only common sense and good laboratory skills. First, consider that every surface you touch and the air that you breathe may be contaminated by microorganisms. Then think about the steps you can take to minimize your exposure to unwanted invisible intruders. You should also be thinking about how to prevent contamination of your bacterial cultures with bacteria from the surrounding environment (which includes you).

To maintain an aseptic work environment, everything you work with should be initially free of microbes. Thus, we begin with pre-sterilized pipettes, culture tubes, and glassware. Inoculating loops and needles made of metal wire can be used to transfer bacteria from one medium to another, such as from the surface of an agar plate to a broth. Metal tools may be sterilized by heating them in the flame of a Bunsen burner. Glass tools or metal spreaders or forceps that can't be sterilized by direct heat are dipped in alcohol followed by a brief pass through the flame to speed the evaporation process. Standard aseptic techniques used for culturing bacteria will be demonstrated at the beginning of lab.

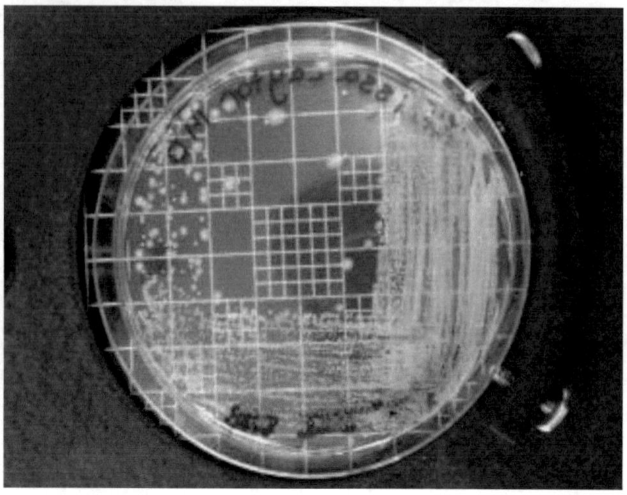

Figure 2. Colonies on an agar plate

One very important method in microbiology is to isolate a single type of bacteria from a source that contains many. The most effective way to do this is the **streak plate** method, which dilutes the individual cells by spreading them over the surface of an agar plate (see Figure 2). Single cells reproduce and create millions of clones, which all pile up on top of the original cell. The piles of bacterial cells observed after an incubation period are called **colonies.** Each colony represents the descendants of a single bacterial cell, and therefore, all of the cells in the colonies are clones. Therefore, when you transfer a single colony from the streak plate to new media, you have achieved a pure culture with only one type of bacteria.

Different bacteria give rise to colonies that may be quite distinct to the bacterial species that created it. Therefore, a useful preliminary step in identifying bacteria is to examine a characteristic called **colonial morphology**, which is defined as the appearance of the colonies on an agar plate or slant. Ideally, these determinations should be made by looking at a single colony; however, if the colonial growth is more abundant and single colonies are absent, it is still possible to describe some of the colonial characteristics, such as the texture and color of the bacterial growth.

Describing Colonial Morphology of Bacteria

By looking closely at the colonial growth on the surface of a solid medium, characteristics such as surface texture, transparency, and the color or hue of the growth can be described. The following three characteristics are readily apparent whether you're looking at a single bacterial colony or more dense growth, without the aid of any type of magnifying device.

Texture—describes how the surface of the colony appears. Common terms used to describe texture may include smooth, glistening, mucoid, slimy, dry, powdery, flaky etc.

Transparency—colonies may be transparent (you can see through them), translucent (light passes through them), or opaque (solid-appearing).

Color or Pigmentation—many bacteria produce intracellular pigments which cause their colonies to appear a distinct color, such as yellow, pink, purple or red. Many bacteria do not produce any pigment and appear white or gray.

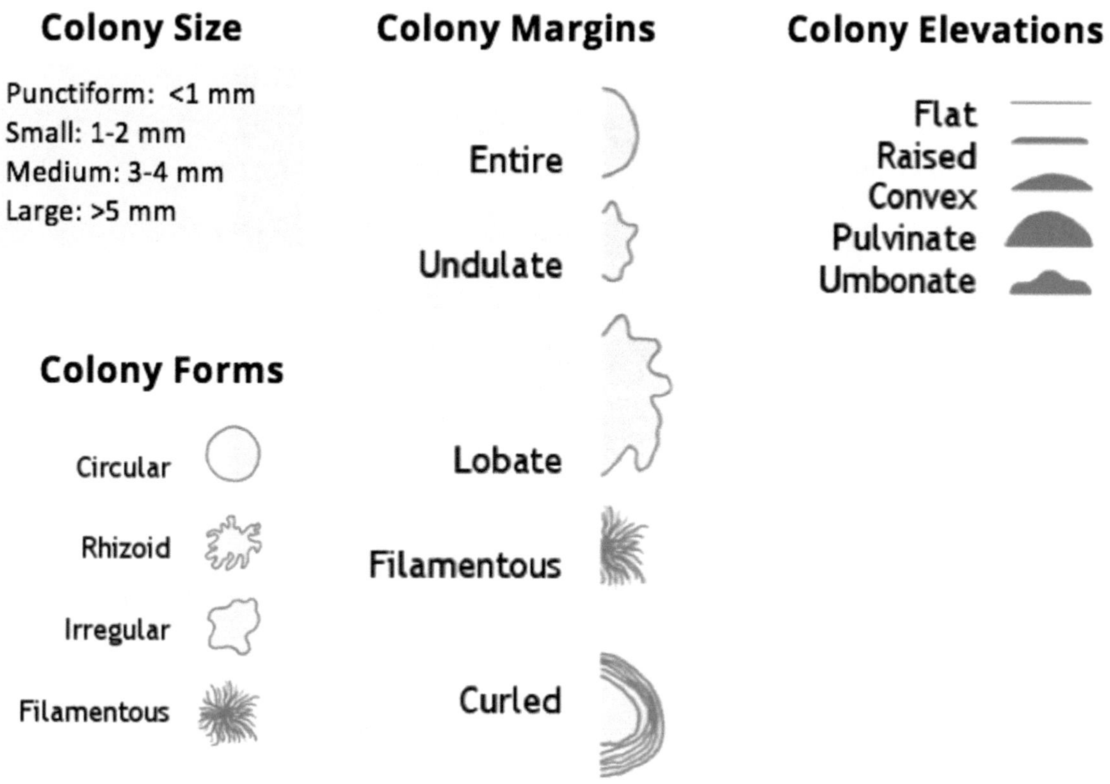

Figure 3. Bacteriological descriptions of colonial morphology

As the bacterial population increases in number, the colonies get larger and begin to take on a shape or form. These can be quite distinctive and provide a good way to tell colonies apart when they are similar in color or texture. The following three characteristics can be described for bacteria when a single, separate colony can be observed. It may be helpful to use a magnifying tool, such as a colony counter or dissecting microscope, to enable a close-up view of the colonies. Colonies should be described as to their overall size, their shape or form, what a close-up of the edges of the colony looks like (edge or margin of the colony), and how the colony appears when you observe it from the side (elevation).

Figure 4 shows a close-up of colonies growing on the surface of an agar plate. In this example, the differences between the two bacteria are obvious, because each has a distinctive colonial morphology.

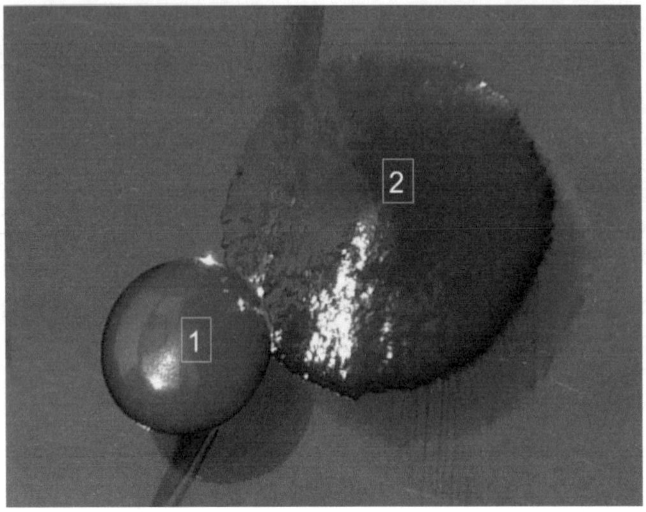

Figure 4. Two different types of bacterial colonies on an agar plate.

Using microbiology terms, describe fully the colonial morphology of the two colonies shown above. A full description will include texture, transparency, color, and form (size, overall shape, margin, and elevation).

Colony 1	
Colony 2	

Now describe the colonial morphology of *Micrococcus luteus*, using the TSA plate culture of this bacterium provided to your group at the beginning of lab:

Size: ____

Texture: ____

Transparency: _______________________________________

Pigmentation: _______________________________________

Form (shape, margin, elevation): _______________________

Media Considerations

A culture medium must contain adequate nutrients to support bacterial growth. Minimally, this would include organic compounds that can provide the building blocks necessary for cellular reproduction. In many cases, predigested protein, such as hydrolyzed soy protein, serves this purpose and will support the growth of many different bacteria. These media formulations are generally referred to as **complex media**, to indicate that it is a mixture with many components.

Many media contain additional substances such as an antibiotic that may be **selective** for a particular type of bacteria by inhibiting most or all other types. **Differential** media will have additional compounds that permit us to distinguish among bacterial types based on differences in growth patterns. We will eventually use selective and differential media in our experiments, but the focus of this lab is to learn the basic culturing techniques, and therefore, the media used will be Tryptic Soy medium, a complex medium formulated with hydrolyzed soy protein.

The media you use in this lab and in all of the future labs will have already been prepared, but it is important for you as a budding microbiologist to understand and appreciate how culture media is prepared. With this in mind, your instructor may have you watch a brief video that demonstrates the art of media making.

Liquid media

Pre-sterilized glass or plastic graduated pipettes (Figure 5) are used to transfer specific volumes of sterile liquids accurately. It is important that you learn how to use these tools correctly, since it may be necessary to transfer sterile and sometimes contaminated liquids among various bottles and tubes. Their appropriate use will be discussed and demonstrated in lab. Some tips to remember:

- The pipette and the media are sterile; there should never be any direct contact with your hands, skin, or lab surfaces.
- Caps or lids on tubes or bottles should never be set down on lab surfaces.
- Tubes or bottles should be held at an angle during the transfer process, to minimize the potential for airborne contaminants to make their way into the opening.
- Passing the opening of the tube or bottle briefly through a flame before and after the transfer process will discourage airborne contaminants from getting into the sterile liquid.

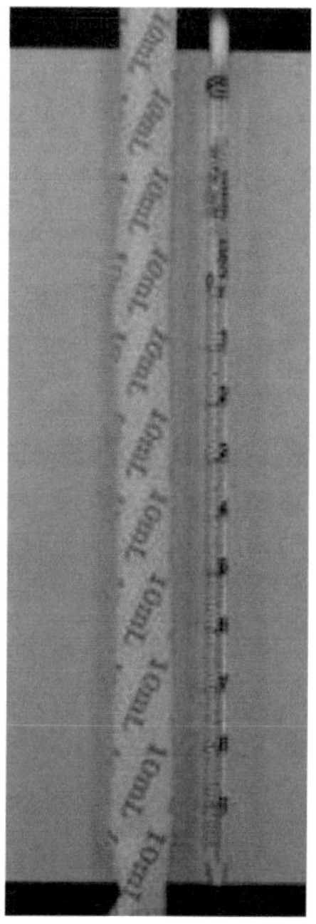

Figure 5. Graduated pipettes for transfer of liquids

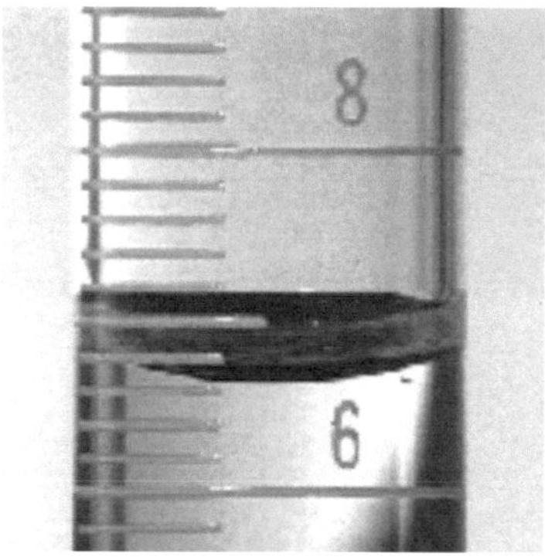

Figure 6. Measuring volume

Pipette practice:

- Obtain water in a small beaker, a 10 ml sterile graduated pipette, and a pipette aid (pipump). Take a minute to note the divisions on the pipette and to understand what volume each mark represents. Use of the pipette to transfer liquids will be demonstrated. Before trying to pipette a sterile liquid, practice drawing up 5 ml of water from the beaker, and releasing it back into the beaker in 1 ml increments. Continue until you feel comfortable holding the pipette and using the pipump. Then practice it again with water in a capped media bottle using aseptic techniques.

A portion of a 10 mL graduated pipette is shown in Figure 6. What is the volume of liquid in this pipette?

Volume:______________________________

Solid and semi-solid media

Growing cultures of bacteria on solid media (agar plate or slant) permits us to view and identify colonial characteristics, and also provides a way to separate bacteria in a mixed culture. Cultures grown on agar plates usually don't survive for long, since Petri dish lids are not tight fitting and the media (and bacteria) dehydrate. Cultures grown on agar plates should always be handled "bottom-up" to prevent condensation—which often accumulates on the lid of the dish during incubation—from dripping down on the culture.

Bacteria may be grown in agar slant or stab media in tubes if the purpose is to maintain them in a longer term culture. Generally, bacteria grown on slants will remain viable for a few weeks to a few months, and sometimes longer if stored in a refrigerator.

In this laboratory, you will be introduced to aseptic techniques and basic lab skills needed to grow and maintain bacteria in culture. You will be applying these skills often, so mastery is important.

Method

A volunteer from your lab bench should obtain one of each of the following cultures:

- TSA streak plate culture of *Micrococcus luteus* and *Enterococcus faecalis*

What BSL containment level practices should be used?

*M. luteus*_________________

*E. faecalis*_________________

- A "mixed culture" in TSB that contains two different bacteria

Below, write the names of the two bacteria in the mixed culture and the appropriate BSL, as specified by your instructor:

Mixed culture bacterium 1 ____

Mixed culture bacterium 2 ____

The techniques needed will first be demonstrated by your instructor. After the demonstration, perform the following tasks, and record your observations/results.

Broth Subculture

Obtain 2 sterile glass culture tubes, a bottle of Tryptic Soy Broth (TSB) and a test tube rack. With small pieces of colored tape, label each tube with your name and either "S" for subculture, or "C" for control. Using aseptic technique, use a 10 ml graduated pipette to transfer 2 ml of broth to each tube.

As demonstrated, use a flame-sterilized inoculating loop to pick up from the surface of the *M. luteus* streak plate culture, a single colony (if small) or a part of a colony (if large) and transfer it to the broth in the tube labeled "S." Add nothing to the second tube "C" which will serve as a sterility control.

Note how the broths look immediately after you inoculate them (they should still look mostly clear). Bacterial growth in broths is indicated by the development of a cloudy appearance. If the newly inoculated broth looks cloudy at the start, you will have no way to determine if this is due to bacterial growth during the incubation period. If your broth looks cloudy, discard it and make another broth using less bacteria.

Place the broth subcultures in an incubator at the temperature and time specified by your instructor.

Streak Plate

Separation of a mixed culture into individual colonies that can be subcultured to make pure cultures depends on how well the streak plate is prepared. The goal of streak plate method is to dilute the cells by spreading them out over the surface of the agar. This is accomplished in stages, as will be demonstrated in lab before you try it yourself.

Use the simulated agar surface below to practice the streak pattern using a pen or pencil.

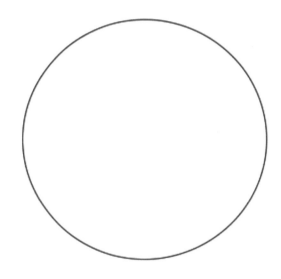

Obtain two TSA plates, and write your name on the bottom half (the half containing the media) around the edge and following the curve (so the writing won't hide your view of the bacterial colonies once they grow). Also write *M. luteus* on one plate (the name of the bacteria you will subculture to this plate). On the other, write "mixed" to indicate that you're subculturing from the mixed culture broth to this plate.

As demonstrated, use a sterilized inoculating loop to pick up one *M. luteus* colony (or a piece of a colony) and transfer it to the surface of the agar plate. Spread the bacteria over approximately a quarter of the plate, edge to edge. Consider this step 1.

Flame the loop and cool it in the agar. Overlap the step 1 streak 3-4 times to pull out a reduced number of bacteria, and spread them out down the side of the plate. Consider this step 2.

Flame the loop and cool it in the agar. Overlap the step 2 streak 3-4 times and spread over the surface. Continue this process, flaming the loop in between each step, until the entire surface of the agar plate is covered.

After performing this with the *M. luteus* culture for practice, repeat the process with a drop of the mixed culture broth that you transfer to the plate with a sterile inoculating loop.

Place the streak plate subcultures in an incubator at the temperature and time specified by your instructor.

Slant Subculture of *M. luteus*

Obtain one slant tube containing TSA, and label it using a small piece of tape with your name and culture name (*M. luteus*). Using a sterilized inoculating **loop**, pick up a bacterial colony (or piece of a colony) from the surface of the plate culture of *M. luteus*, and inoculate the surface of the slant. Place the slant subculture in an incubator at the temperature and time specified by your instructor.

Stab or Deep Tube Subculture of *E. faecalis*

Obtain one stab tube containing semisolid TSA, and label it using a small piece of tape with your name and culture name (*E. faecalis*). Using a sterilized inoculating **needle**, pick up a bacterial colony (or piece of a colony) from the surface of the plate culture of *E. faecalis*, and inoculate the media by stabbing the needle into the center of the agar in the tube,and pushing it down to the bottom. Withdraw the needle carefully and try to remove it by following the same stab line that you made pushing the needle down. Place the stab subculture in an incubator at the temperature and time specified by your instructor.

A note about incubation temperatures

As you will learn, bacteria have preferred growth temperatures where their reproduction rate is the greatest. All of the bacteria we work with in lab are **mesophilic**, which means that they grow at temperatures between 20–40°C. However, some prefer body temperature (37°C), while others grow best at room temperature (approximately 25°C). This lab is equipped with incubators set at either temperature.

How long you plan to leave your cultures in an incubator should also be a consideration. Growing cultures at the higher temperature may speed their rate of growth, but it also causes dehydration of the media and an earlier demise to the bacteria in the culture.

As a general rule, **for bacteria that grow best at body temperature**, if you intend on returning to lab within 24 to 36 hours (highly recommended), then incubate them at 37°C. If you cannot return to lab during an "open lab" period, then incubate them at room temperature, or arrange to have your cultures transferred to a refrigerator after they grow, so that the culture won't die out before you can finish your experiments. Bacteria that grow best at room temperature should always be incubated at room temperature, and growth may take a little longer.

Primary culture from an environmental source—you!

With your introduction to basic bacteriological culturing techniques complete, it's time to apply those skills. Today is the beginning of **The Human Skin Microbiome Project**, which starts with the primary culture of bacteria from your skin on TSA medium. It is important that you read the project description (in the next chapter) so that you understand the goals and the scope of the project.

To begin, you will take a sample from your skin. Your first decision will be what part of your skin do you want to sample? Note: ONLY external skin surfaces are permitted.

Obtain a sterile swab and a tube of sterile distilled water, and label a TSA plate with your name and the date. Remove the wrapping from the swab and soak it in sterile water, using aseptic technique.

Rub the wet swab back and forth firmly over the area of skin you have chosen to sample. Then rub the swab over approximately a third of the surface of the TSA agar plate.

Sterilize an inoculating loop, and complete the rest of the streak plate pattern using the loop. Incubate this plate at room temperature for up to a week.

After incubation, look to see if isolated colonies have developed on the plate. If there are no colonies or no isolated colonies, you will need to make another streak plate with the advice of your instructor on how to proceed. If there are isolated colonies, transfer the plate to the refrigerator. From this plate you will ultimately choose one single colony and prepare a pure culture. The criteria for colony selection and next steps are described in the next chapter, "The Human Skin Microbiome Project."

To complete the lab, the bacteria in the cultures have to grow. Therefore, the following observations are made AFTER the cultures have had time to grow.

Observations and Outcomes

Broth subcultures

Look at the broth subculture tubes, and describe what you expected to see, and how they appear in terms of how "cloudy" they look—cloudiness is an indication of bacterial growth.

	Cloudiness of broth before incubation	*Predicted* appearance of broth after incubation	*Actual* appearance of broth after incubation
M. luteus subculture ("S")			
Sterility Control ("C")			

Streak plate subcultures

Look at the streak plate subcultures that you made. Conduct a self-assessment of how well you performed the technique. What you hope to see are individual colonies, well separated from each other. On the streak plate of the mixed culture, you should be able to see two distinctly different types of colonies.

***M. luteus* streak plate:**

- Are the colonies well separated?

- How many different types of colonies do you see?

- Describe in full the colonial morphology of the bacteria on this plate:

Streak plate of the mixed culture:

- Are the colonies well separated?

- Could you make a pure culture of both bacteria from this plate? If you think you can, subculture a single colony of each type to one half of a TSA plate, divided by drawing a line with a marker on the bottom of the plate, as shown below. Incubate the plate, and then observe to see if you successfully separated the two bacteria in the mixed culture into two pure cultures. Use this self-analysis to consider improvements you might make in the technique you applied to making the streak plate.

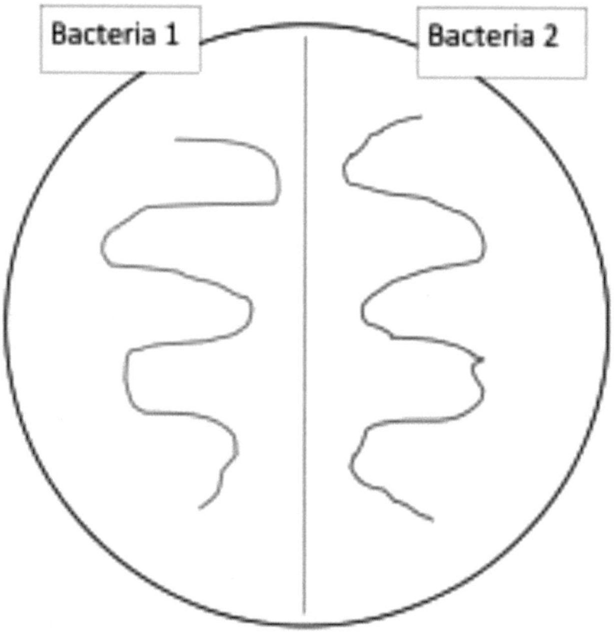

- Describe in full the colonial morphology of both of the bacteria from the mixed culture:

	Colony Type 1	Colony Type 2
Size		
Texture		
Transparency		
Pigmentation		
Whole Colony		

TSA slant subculture

Examine the subculture of *M. luteus* you prepared on the TSA slant.

- Describe the **texture, transparency, and pigmentation** of the bacterial growth on the slant. Only these characteristics can be described for a slant culture, since there should be no discreet colonies on the slant, only an area of dense growth along the streak line.

- Does your description match what was noted for the *M. luteus* colonies when you described the colonial morphology previously?

- Do you see evidence of any other type of bacteria (meaning a different colonial morphology) on the slant?

- Is this a pure culture?

TSA stab subculture

Look closely along the stab line in the media in the tube. Do you see evidence of bacterial growth? If yes, describe and/or sketch how it appears.

Semisolid agar of the type used in this exercise can be used as a way to evaluate if a bacteria is motile, meaning in possession of one or more flagella that facilitates movement through liquids or semisolids. The way to evaluate motility is to look closely at the line of inoculation you created when the tube was stabbed. Nonmotile bacteria will grow along the stab line only. If they are motile, they will be able to move through the semisolid agar (like swimming through jello), and you won't be able to see a distinct line in the agar—just cloudiness surrounding the stab line.

- Based on your observation of the bacteria in the stab culture, is there evidence that the bacteria are motile?

- For bacteria, the ability to move (motility) requires that they have which specific cellular structure?

The Human Skin Microbiome Project

Project Goals and Objectives

The Human Skin Microbiome Project is an application of the principles and practices of classic microbiological investigation. During this project, which will take several weeks to complete, you will survey the microbial inhabitants of the human skin (yours). Then, you will apply an expanding array of microbiology laboratory skills to grow and investigate the colonial, cellular, and metabolic properties of one of the bacterial species from your skin culture.

More specifically, you will:

- prepare a primary culture from your skin, and observe the colonial and cellular properties of the bacteria that grow on it;
- identify one skin isolate that you would like to investigate further, and maintain it in a pure culture for an extended period of time;
- use microbiological methods to investigate the cellular and metabolic properties of your skin isolate;
- understand basic principles of taxonomy and how to apply the information in *Bergey's Manual of Determinative Bacteriology* to presumptively identify your skin isolate.

Biosafety Considerations

While it may seem somewhat ironic that bacteria you have been carrying around on your skin forever are now going to be classified as a potential biohazard and subjected to risk assessment and laboratory containment practices, it is nonetheless an important consideration.

Of the various types of bacteria that might be encountered during the primary culture stage of this project, most are BSL-1, and some BSL-2. To minimize the risks of working with bacteria of unknown identity, this project will be limited to only **Gram positive bacteria**, and **BSL-2 practices** will be employed while working with cultures of your isolate.

Bergey's Manual

The definitive reference book for bacterial taxonomy and identification is *Bergey's Manual*. There are two versions: the manual of *Systematic Bacteriology*, which is concerned with issues of bacterial taxonomy and systematics (arranging bacteria into taxa according to similarities/differences in DNA), and the manual *of Determinative Bacteriology,* which deals specifically with the identification aspects of bacterial taxonomy. The latter book will be our primary resource for this project. For an overview of the manual and how to use it, **read Chapters I, II, III, IV, and V.**

The Human Skin Microbiome

The bacteria and other microbes that live on human skin are those that are best adapted to survive the prevailing conditions. Regions of the human body can be thought of as different ecosystems. Exposed, dry areas of the skin, such as the forearm, are akin to an arid desert environment, which is a preferred environment for many Gram-positive bacteria. Skin sites that are generally dark, warm, and moist, such as the underarm or perineum, are similar to temperate or tropical ecosystems; they tend to harbor more microbes in general and are more likely to have a larger percentage of Gram-negative bacteria.

The quantitative differences found at these sites may relate to the amount of moisture, body temperature, and varying concentrations of skin surface lipids. Most microorganisms live in the superficial layers of the skin (the stratum corneum) and in the upper parts of the hair follicles. Some bacteria, however, reside in the deeper areas of the hair follicles, where they may be beyond the reach of ordinary disinfection procedures (like washing your face with soap/water or an antibacterial product). These out-of-reach bacteria serve as a reservoir for recolonization of the skin environment after the surface bacteria are removed.

Figure 1 illustrates the types of bacteria that are commonly found on various regions of the human skin. Not all of these bacteria are culturable, because the growth conditions necessary for their survival are difficult to replicate in an artificial environment. Using the culture conditions established for this laboratory, the bacteria grown in your primary culture will most likely be Actinobacteria (*Micrococcus, Corynebacterium, Mycobacterium*), Firmicutes (*Staphylococcus* or other Gram-positive bacteria), or Proteobacteria (Gram negative bacteria).

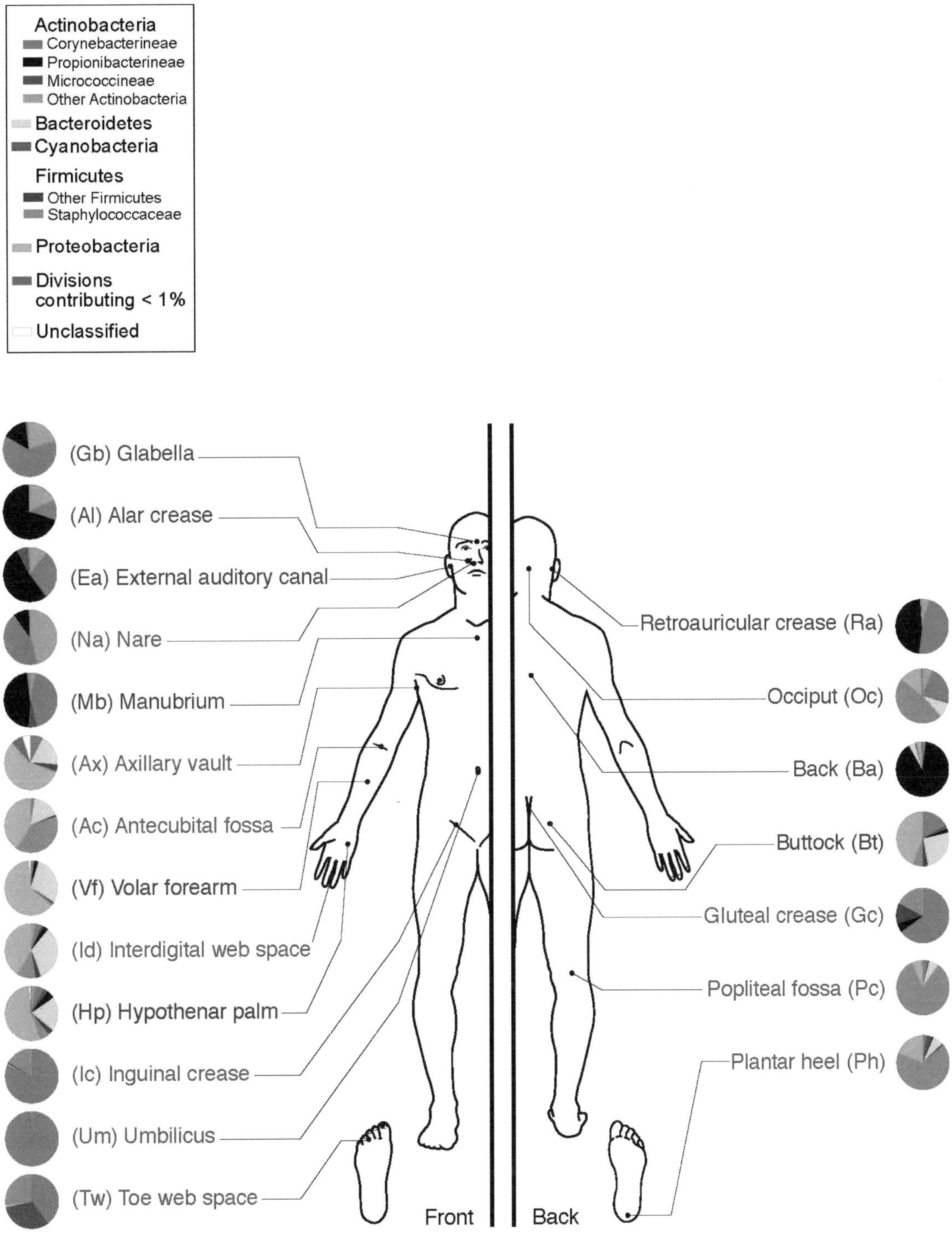

Figure 1. Types of bacteria found on human skin

The skin microbiome may include fungi such as yeasts and molds as well as bacteria. While interesting, these eukaryotic microorganisms are outside of the scope of this project. Molds form very distinctive colonies that will be easy to identify as fungal in origin, and thus, easy to avoid. Yeasts, however, produce colonies that resemble those of bacteria, although typically smaller and different in color. When you select the colony and make a pure culture, avoid colonies that have the appearance of either a mold (furry, fuzzy, or powdery) or a yeast (very small, very slow growing—only appearing after a week or more of incubation, and brightly colored—red, orange, pink, or even bright white colonies).

Over the course of several weeks, you will maintain your bacterial strain in pure culture while performing tests to determine its colonial, cellular, and metabolic properties, and ultimately its "presumptive" identity. The term "presumptive" is used because phenotypic methods are less exact than those that rely on a direct analysis of DNA.

Record all observations, test outcomes, and interpretations on the Human Skin Microbiome Project Worksheets, according to the instructions provided by your instructor.

Identification of an Unknown Bacterium

Bergey's Manual contains an enormous amount of information about the characteristics of all known bacterial species, mostly presented in table form. You will use the information in these tables to determine the identity of your skin isolate. To aid in this process, and as a way to demonstrate how you ruled out all other possible bacterial species, a taxonomic tool called a dichotomous key (also called a diagnostic key or sequential key) will be used to narrow down the possibilities.

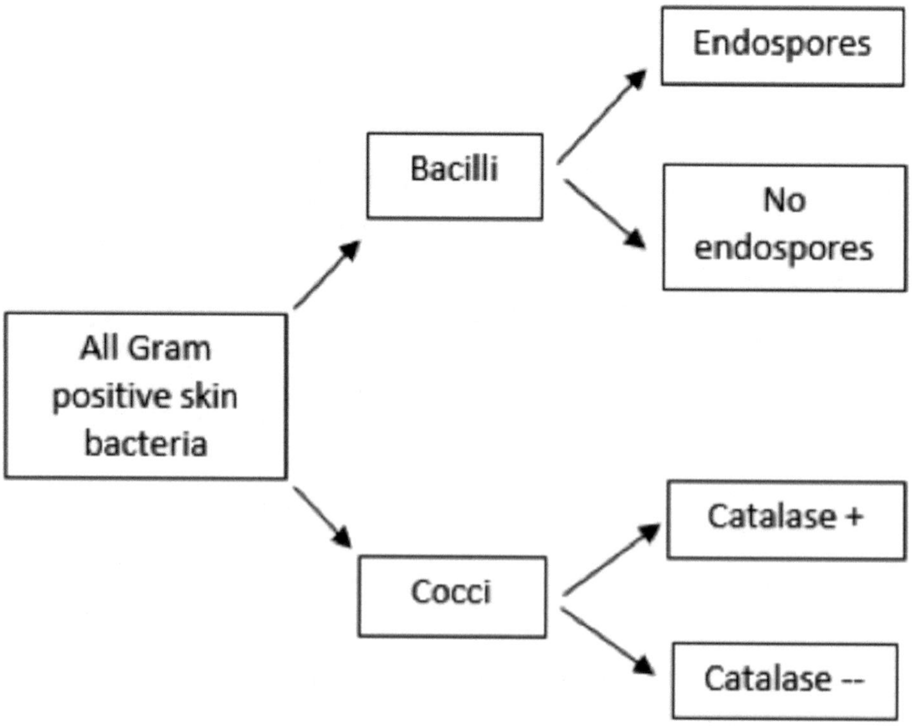

Figure 2. Example of a dichotomous key

A dichotomous key is a sequence of questions with two possible mutually exclusive answers (a "couplet," such as yes or no, positive or negative, cocci or bacilli). Starting with a large group of bacteria (in this case, all of the possible Gram-positive bacteria that can be found on the human skin), the first question should relate to an observable characteristic for which there are two choices, followed by additional questions until the possibilities are narrowed to a single choice. A brief example is shown in Figure 2.

Bergey's Manual of Determinative Bacteriology is designed to facilitate identification, and not classification, of bacteria based primarily on phenotypic observations. In Bergey's, the broadest grouping is represented by the four **Major Categories.** The primary criterion to establish the Major Category for your skin microbe will be the nature of its cell wall. Remember that you are being limited to choosing a bacterium that is Gram positive; therefore, the starting point for this project will be bacteria that are in **Major Category II: Gram-positive eubacteria that have cell walls.** Because of the culture methods and conditions we will employ in this project, it is highly unlikely that your isolate would be from Categories III (eubacteria lacking cell walls) or IV (archaeobacteria); thus, those are excluded.

At this point, you have narrowed your options down from all bacteria to only those classified as Category II. The next step is to identify the **Group**, within the Major Category, to which your isolate belongs. Chapter V in Bergey's provides a list and brief summary of the bacteria in the Groups within each Category; Table V.2 refers specifically to Category II bacteria. Cellular morphology (for example, cell shape and presence of endospores) and physiology (particularly with regard to your isolate's physiological oxygen requirements and relative results) will help you decide on a Group. Once you made that decision, you can follow the directions given in Table V.2 to find the first page of the identification tables in Bergey's.

Continue to narrow your choices down to a single **Genus** within the Group, and once you've decided on a genus, locate and refer to the identification table that specifies the different **Species** within that genus. The tables also list the laboratory tests routinely performed to differentiate among the species, and the expected outcomes of those tests for each bacterial species in the genus.

You may not be able to conduct all of the tests listed in *Bergey's*. Different tests may be available for you to perform, depending on the laboratory in which you work. For the purpose of identifying your skin microbe, below is a list of the tests that are available for you to use in the ID process.

- Gram stain and endospore stain
- 7.5% salt tolerance (MSA)
- Mannitol fermentation (MSA for salt-tolerant bacteria)
- Bile tolerance and esculin hydrolysis (BE)
- Hemolysis on blood agar
- Glucose fermentation (MR; VP; TSI)
- Lactose/sucrose fermentation (TSI)
- Fermentation of other carbohydrates (see instructor)
- Susceptibility or resistance to antibiotics
- Acetoin production (VP)
- Catalase

- Oxidase
- Production of H_2S (TSI; SIM)
- Indole (SIM)
- Motility (SIM)
- Citrate utilization
- Nitrate reduction
- Coagulase (for staphylococci ONLY)
- Urease

Other tests may be available, as indicated by your instructor.

Observations, Outcomes, and Next Steps

The following tasks will be performed over the course of several weeks:

Project Step 1: Make a primary culture from your skin on a TSA plate and incubate it for up to a week at room temperature.

Project Step 2: After incubation, the TSA plate that contains the primary culture from your skin will very likely hold many, hopefully well isolated, colonies.

From among these, choose 3 colonies for subculture and further examination. Remember to avoid colonies that appear to be a mold (typically large, green or brown, and fuzzy) or yeast (**very** small and vibrantly white or a shade of red). With a sterile inoculation loop, subculture each colony to a section of a TSA plate, to create a pure culture, as you did previously and illustrated in Figure 3. Incubate the plate at RT until bacterial growth is abundant.

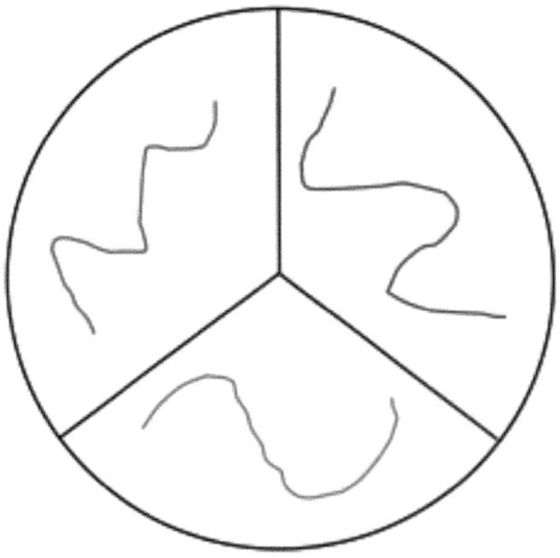

Figure 3. Subcultures on different sections of a TSA plate

Project Step 3: Gram stain each of the three pure cultures. Choose one that is Gram-positive (bacilli or cocci) as your project bacterium (your skin microbe, or HSM). Subculture your HSM to a TSA slant and incubate it. Once there is abundant growth on the surface, store the TSA slant culture in the refrigerator. Make sure the slant is clearly labelled with your name.

Complete the Colonial and Cellular Morphology (CCM) Worksheet.

Project Step 4: From the observations you've made and the results of lab tests,use Bergey's Manual as a reference to determine the Major Category, Group and then Genus of your HSM.

Complete the Genus Identification (GID) Worksheet.

Project Step 5: Once you've narrowed your options down to a single genus, locate the specific table in Bergey's Manual for identification of individual species. Based on the information in the table, compile a list of appropriate tests that will facilitate species identification. Select those that are available for you to perform, as specified by your lab instructor. Perform the appropriate tests and record the results.

Project Step 6: Compare your results with the expected test outcomes from Bergey's Manual to determine the Species of your human skin microbe.

Complete the Species Identification (SID) Worksheet.

HSMP Worksheet 1: Colonial and Cellular Morphology

Name ____ Date due ________________

Record the observations/results obtained so far below. NOTE: your instructor may make this worksheet available to you electronically through a course management system, and may request that you type your answers into the worksheet before printing and handing it in.

1. Name the region of your body from which you obtained the specimen for the primary culture.

2. Based on colony appearance, approximately how many different types of bacteria from your skin are represented on the TSA streak plate of your primary culture?

3. Based on the appearance of an **isolated colony** and using appropriate microbiology terminology, describe the colonial morphology of each of the three bacterial subcultures.

Colony Type 1	
Colony size	
Texture	
Transparency	
Pigmentation	
Form (shape, margin, elevation)	
Colony Type 2	
Colony size	
Texture	
Transparency	
Pigmentation	
Form (shape, margin, elevation)	
Colony Type 3	
Colony size	
Texture	
Transparency	
Pigmentation	
Form (shape, margin, elevation)	

4. For each of the three bacterial pure cultures, describe the outcome of the Gram stain using appropriate microbiology terminology:

	Gram Stain Outcome	Cell shape	Arrangement
Colony 1			
Colony 2			
Colony 3			

5. Of the three bacteria you investigated, choose one that is Gram-positive as your project bacterium. Below, indicate which of the three you chose and restate the Gram staining result.

Colony # and Gram stain result (including cell shape and arrangement):

____

6. Compare and contrast the chemical composition and structure of the cell wall of a Gram-positive bacterium such as your isolate, with the cell wall of a Gram negative bacterium.

7. Briefly discuss why Gram-positive cells appear purple, and Gram-negative cells appear pink, after the Gram stain process is applied.

8. Briefly discuss how bacterial cells produce "arrangements" that we can observe with a microscope.

9. State whether it will be necessary for you to perform an endospore stain on your isolate, and give a specific reason to explain why you should, or should not, use this staining method to identify your isolate.

10. Briefly explain why it is necessary to include a mixture of iodine and potassium iodide (Gram's Iodine) in the overall Gram stain procedure.

In addition to this worksheet, you may also be asked to prepare and provide to your instructor a **Gram stained slide** of a smear prepared from your environmental isolate to evaluate your technique. If evaluated, the following criteria will be used: single layer of cells is achieved, cell morphology and arrangement are easily determined, all cells appear the same color, shape, and arrangement, and there are no visible contaminants.

Instructor Evaluation of Gram Stained Slide:

Gram stain result as observed by instructor: ______________________________

Evaluation of technique: ____

Criteria: Single layer of cells is achieved; cell morphology and arrangement are easily determined; all cells appear the same color, shape, and arrangement; and there are no visible contaminants. Degree of concurrence with instructor's description of cellular morphology will also be noted.

HSMP Worksheet 2: Genus Identification

Name ______________________________________ Date due _______________

1. From your CCM worksheet, restate the Gram stain results (reaction, morphology, and arrangement) for your skin isolate:

2. Taxonomic Classification: State the **Domain** and **Phylum** for your isolate, based on the evidence you have accumulated so far.

3. Report the results of the following physiological tests performed on your environmental isolate:

Test	Describe **in detail** the outcomes of the following tests (meaning, what you directly observe: such as bubbles after H_2O_2 was added; red color on the slant and yellow on the butt with cracks; etc.)	Interpretation of observed outcome (for example; pos or neg; K/A, gas, etc)
Catalase		
Oxidase		
TSI agar		
Nitrate Reduction		

4. From your observations of bacterial growth characteristics and physiological tests up to this point, state ALL energy metabolism pathway(s) used by your skin bacterium to make ATP. Then provide convincing evidence from among your observations and test results to support your determination.

Energy Metabolism pathway	Do your observations and/or results of the tests above indicate that your EI uses this pathway? (**YES** or **NO**)	STATE one observation and/or test result that provides scientific evidence for that pathway, and explain why/how the result indicates that your EI bacterium uses this pathway.
Aerobic respiration		
Anaerobic respiration		
Fermentation		

5. Growth Category for Oxygen

Based on the observed growth patterns and test results above, state the physiological oxygen requirement (strict aerobe, microaerophile, strict anaerobe, facultative anaerobe, or aerotolerant anaerobe) for your EI bacterium.

6. Bergey's Group

Table V in Bergey's is divided into four sections, one for each Major Category of bacteria. Remember that you were directed to select a Gram-positive bacterium as your EI, so look at Table V.2 to determine to which Group, within Major Category II, your EI should be assigned. Based on the characteristics of your EI bacterium that you have observed up to this point:

(a) State the **Group** (group **number** and **name**) for your EI according to the Bergey's Manual identification system.

(b) State TWO observations that provide **scientific evidence** to support your choice of Group designation from (a) above.

7. Tests to assign Genus

If necessary (**and it may not be necessary at this point**), perform additional tests to determine the genus of your environmental isolate. Describe those tests and their outcomes in the table below. If you can determine the genus without additional tests, don't put anything in this table.

Test/Observation	Describe in detail the outcomes of the following tests (meaning, what you directly observe): such as bubbles after H_2O_2 was added; red color on the slant and yellow on the butt with cracks; etc.)	Interpretation of observed outcome (for example; pos or neg; K/A, gas, etc)

8. Genus ID Flowchart (dichotomous key)

A brief example of how to construct a dichotomous key was provided previously in this lab. Note that the key you develop will be used by your instructor to review the process and logic of your choice of genus.

Some advice on how to proceed: Remember that your goal is to **RULE OUT** genera that are not consistent with the characteristics you've observed for your EI bacterium. Start by listing **ALL** of the possible genera in the Bergey's Group. Then look at the characteristics that distinguish the various genera from one another (such as cell shape, endospore production, growth on human skin, etc.). As your first couplet, choose a feature that your isolate exhibits and the majority of other genera lack. Then, for the genera that were not ruled out, choose as the next couplet a feature that again rules out as many genera as possible. Continue until there is only a single genus left that is consistent with all of your observations made up to this point.

State the **presumptive genus** of your EI bacterium: ________________________________

HSMP Worksheet 3: Species Identification

Name ______________________________Date Due ____________

1. Review of observations/characteristics of your EI determined so far:

Pigmentation of colonies (color ONLY)	
Gram stain reaction	
Cellular morphology	
Cellular arrangement	
Endospores observed (yes or no)	
Result of catalase test (+ or -)	
Result of the oxidase test (+ or -)	
Result of TSI	
Result of nitrate reduction test (+ or -)	
List ALL energy pathways indicated	
Growth category for oxygen	
Bergey's Group (**# and name**)	
Name of Genus	

2. Tests for Species ID

Locate the specific Table in Bergey's Manual that shows the species within the genus and the tests needed for the differentiation and identification of your isolate. Below, list the tests you will need to perform (in addition to those already done) to presumptively identify your isolate. Cross reference the list of tests available in your laboratory (provided previously by your instructor).

3. Complete the table below with the tests/outcomes you observed for your EI bacterium. The number of rows in the table is arbitrary – only do as many tests as needed to ID your isolate. Add additional rows, if necessary).

Name of test	Direct observation of the outcome of test(s) (how did the media, slide, tube, etc. APPEAR when you looked at it?)	Outcome/Result (pos or neg, K/A, etc.)

4. As you did previously, construct a dichotomous key to show the process and logic used to presumptively identify your environmental isolate. Begin by listing ALL species within the genus, below. Note that subspecies (if there are any) should be listed individually.

5. **Complete EITHER A or B below:**

A. If you were able to identify a single species after completing all possible tests: Write the full binomial name (using scientific nomenclature) for your environmental isolate below:

B. If you were NOT able to discriminate a single species after completing all your tests: Write the full binomial name of ALL remaining species and provide the reason why you were unable to assign a single species to your EI.

6. Growth Characteristics

Use biology/microbiology terminology to state the specific category related to the following growth characteristics for your EI. Provide your reasoning for the choice of each category, including **specific examples** of growth patterns observed for your isolate from among your observations and tests.

	Physiological Category	**Supporting evidence from among your observations/test results**
Nutritional		
Temperature		
Osmotic (salt) tolerance		

7. Fully **classify** your isolate by providing the following information, using appropriate terminology and scientific nomenclature:

Taxon	**Classification for your skin isolate**
Domain	
Phylum	
Class	
Order	
Family	
Genus	
Species	

Differential Staining Techniques

Viewing Bacterial Cells

The microscope is a very important tool in microbiology, but there are limitations when it comes to using one to observe cells in general and bacterial cells in particular. Two of the most important concerns are resolution and contrast. Resolution is a limitation that we can't do much about, since most bacterial cells are already near the resolution limit of most light microscopes. Contrast, however, can be improved by either using a different type of optical system, such as phase contrast or a differential interference contrast microscope, or by staining the cells (or the background) with a chromogenic dye that not only adds contrast, but gives them a color as well.

There are many different stains and staining procedures used in microbiology. Some involve a single stain and just a few steps, while others use multiple stains and a more complicated procedure. Before you can begin the staining procedure, the cells have to be mounted (smeared) and fixed onto a glass slide.

A bacterial smear is simply that—a small amount of culture spread in a very thin film on the surface of the slide. To prevent the bacteria from washing away during the staining steps, the smear may be chemically or physically "fixed" to the surface of the slide. Heat fixing is an easy and efficient method, and is accomplished by passing the slide briefly through the flame of a Bunsen burner, which causes the biological material to become more or less permanently affixed to the glass surface.

Heat fixed smears are ready for staining. In a simple stain, dyes that are either attracted by charge (a cationic dye such as methylene blue or crystal violet) or repelled by charge (an anionic dye such as eosin or India ink) are added to the smear. Cationic dyes bind the bacterial cells which can be easily observed against the bright background. Anionic dyes are repelled by the cells, and therefore the cells are bright against the stained background. See Figures 1 and 2 for examples of both.

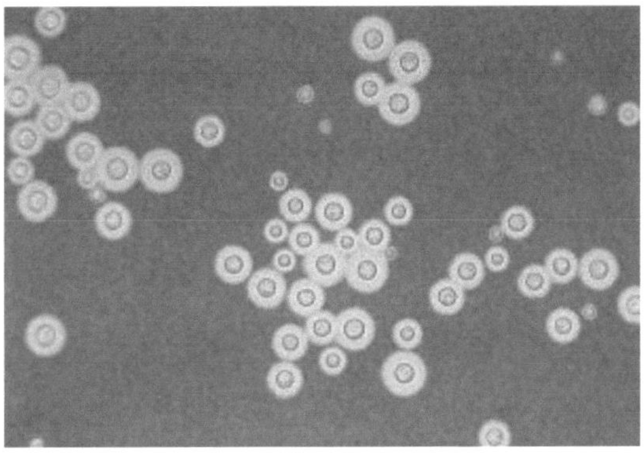

Figure 1. Negative stain of Cyptococcus neoformans, an encapsulated yeast

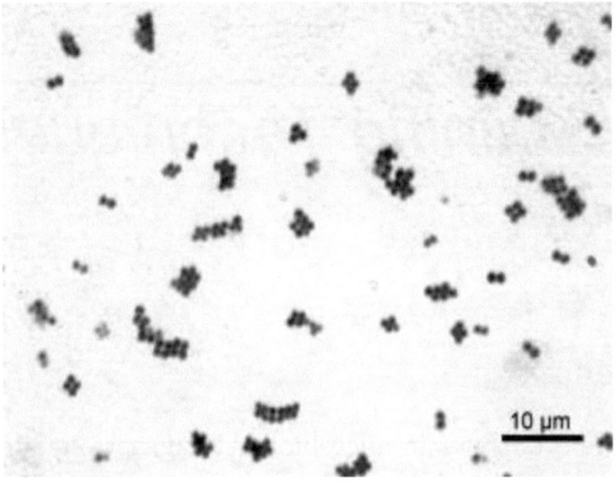

Figure 2. Positive stain of Staphylococcus aureus.

Probably the most important feature made obvious when you stain bacterial cells is their **cellular morphology** (not to be confused with colonial morphology, which is the appearance of bacterial colonies on an agar plate). Most heterotrophic and culturable bacteria come in a few basic shapes: spherical cells (coccus/cocci), rod-shaped cells (bacillus/bacilli), or rod-shaped cells with bends or twists (vibrios and spirilla, respectively). There is greater diversity of shapes among Archaea and other bacteria found in ecosystems other than the human body.

Often bacteria create specific **arrangements** of cells, which form as a result of binary fission by the bacteria as they reproduce. Arrangements are particularly obvious with non-motile bacteria, because the cells tend to stay together after the fission process is complete. Both the shape and arrangement of cells are characteristics that can be used to distinguish among bacteria. The most commonly encountered bacterial shapes (cocci and bacilli) and their possible arrangements are shown in Figures 3 and 4.

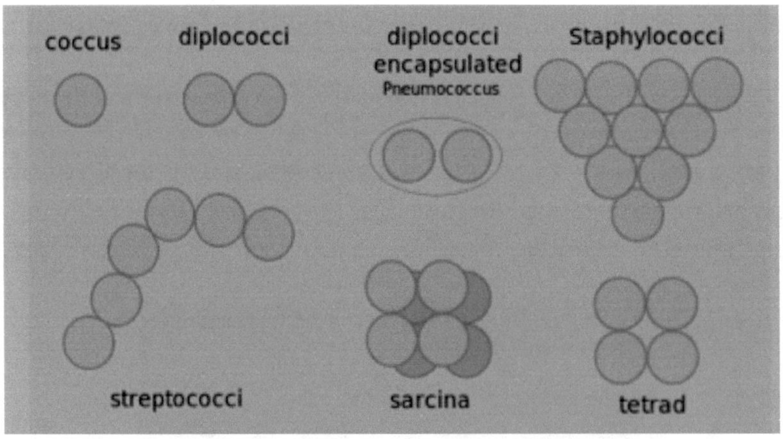

Figure 3. Possible bacterial cell arrangements for cocci

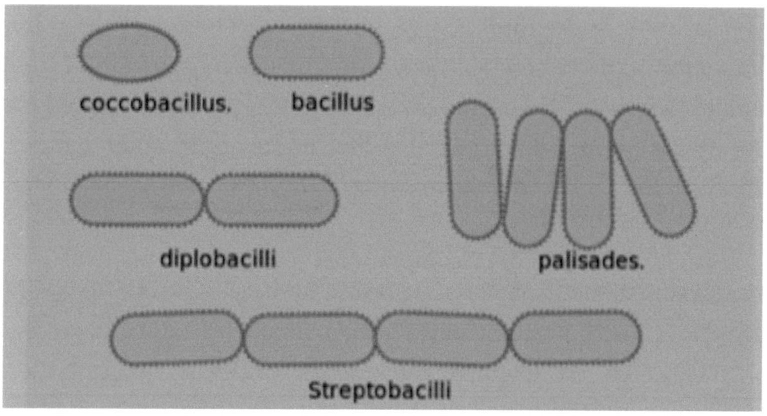

Figure 4. Possible bacteria cell arrangements for bacilli

Differential Staining Techniques

In microbiology, differential staining techniques are used more often than simple stains as a means of gathering information about bacteria. Differential staining methods, which typically require more than one stain and several steps, are referred to as such because they permit the differentiation of cell types or cell structures. The most important of these is the Gram stain. Other differential staining methods include the endospore stain (to identify endospore-forming bacteria), the acid-fast stain (to discriminate *Mycobacterium* species from other bacteria), a metachromatic stain to identify phosphate storage granules, and the capsule stain (to identify encapsulated bacteria). We will be performing the Gram stain and endospore staining procedures in lab, and view prepared slides that highlight some of the other cellular structures present in some bacteria.

Gram Stain

In 1884, physician Hans Christian Gram was studying the etiology (cause) of respiratory diseases such as pneumonia. He developed a staining procedure that allowed him to identify a bacterium in lung tissue taken from deceased patients as the etiologic agent of a fatal type of pneumonia. Although it did little in the way of treatment for the disease, the Gram stain method made it much easier to diagnose the cause of a person's death at autopsy. Today we use Gram's staining techniques to aid in the identification of bacteria, beginning with a preliminary classification into one of two groups: **Gram positive** or **Gram negative**.

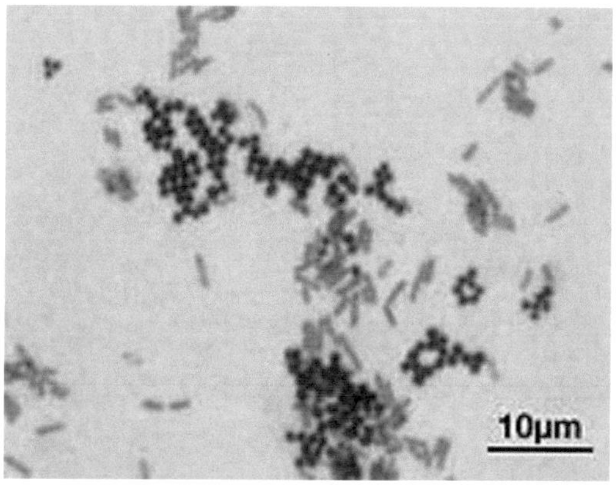

Figure 5. Bacteria stained with Gram stain.

The differential nature of the Gram stain is based on the ability of some bacterial cells to retain a primary stain (crystal violet) by resisting a decolorization process. Gram staining involves four steps. First cells are stained with crystal violet, followed by the addition of a setting agent for the stain (iodine). Then alcohol is applied, which selectively removes the stain from only the Gram negative cells. Finally, a secondary stain, safranin, is added, which counterstains the decolorized cells pink.

Although Gram didn't know it at the time, the main difference between these two types of bacterial cells is their cell walls. Gram negative cell walls have an outer membrane (also called the envelope) that dissolves during the alcohol wash. This permits the crystal violet dye to escape. Only the decolorized cells take up the pink dye safranin, which explains the difference in color between the two types of cells. At the conclusion of the Gram stain procedure, Gram positive cells appear purple, and Gram negative cells appear pink.

When you interpret a Gram stained smear, you should also describe the morphology (shape) of the cells, and their arrangement. In Figure 5, there are two distinct types of bacteria, distinguishable by Gram stain reaction, and also by their shape and arrangement. Below, describe these characteristics for both bacteria:

	Gram positive bacterium:	**Gram negative bacterium:**
Morphology		
Arrangement		

Acid Fast Stain

Some bacteria produce the waxy substance **mycolic acid** when they construct their cell walls. Mycolic acid acts as a barrier, protecting the cells from dehydrating, as well as from phagocytosis by immune system cells in a host. This waxy barrier also prevents stains from penetrating the cell, which is why the Gram stain does not work with mycobacteria such as *Mycobacterium*, which are pathogens of humans and animals. For these bacteria, the **acid-fast staining** technique is used.

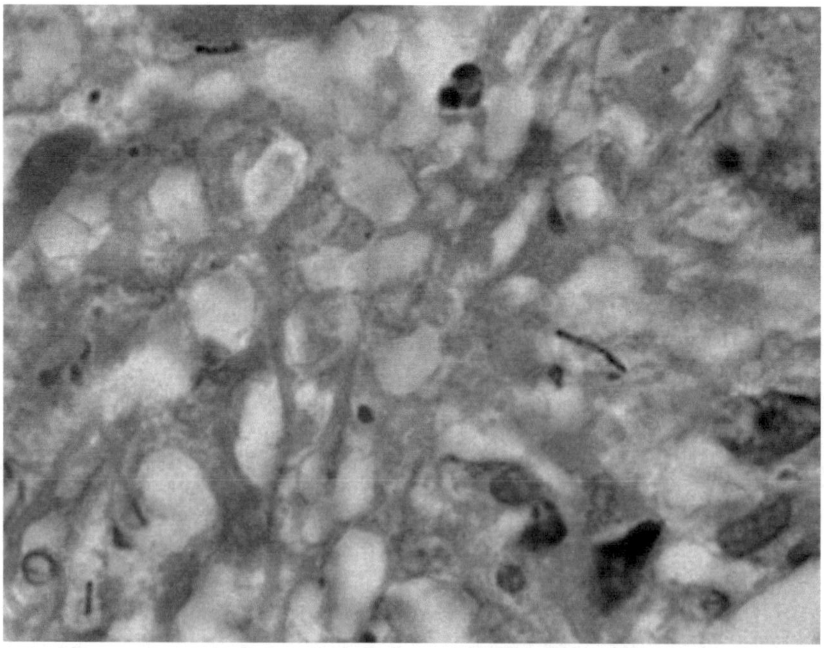

Figure 6. Acid-fast bacilli in sputum

To perform the acid-fast stain, a heat-fixed smear is flooded with the primary stain carbol fuchsin, while the slide is heated over a steaming water bath. The heat "melts" the waxy cell wall and permits the absorption of the dye by the cells. Then the slide is allowed to cool and a solution of acid and alcohol is added as a decolorizer. Cells that are "acid-fast" because of the mycolic acid in their cell wall resist decolorization and retain the primary stain. All other cell types will be decolorized. Methylene blue is then used as a counterstain. In the end, acid-fast bacteria (AFB) will be stained a bright pink color, and all other cell types will appear blue.

Staining Methods to Highlight Specific Cell Structures

Capsule: The polysaccharide goo that surrounds some species of bacteria and a few types of eukaryotic microbes is best visualized when the cells are negative stained. In this method, the bacteria are first mixed with the stain, and then a drop of the mixture is spread across the surface of a slide in the thin film. With this method, capsules appear as a clear layer around the bacterial cells, with the background stained dark.

Metachromatic granules or other intracytoplasmic bodies: Some bacteria may contain storage bodies that can be stained. One example is the Gram positive bacilli *Corynebacterium*, which stores phosphate in structures called "volutin" or metachromatic granules that are housed within the cell membrane. Various staining methods are used to visualize intracytoplasmic bodies in bacteria, which often provide an identification clue when observed in cells.

Endospore Stain

Endospores are dormant forms of living bacteria and should not be confused with reproductive spores produced by fungi. These structures are produced by a few genera of Gram-positive bacteria, almost all bacilli, in response to adverse environmental conditions. Two common bacteria that produce endospores are *Bacillus* or *Clostridum*. Both live primarily in soil and as symbionts of plants and animals, and produce endospores to survive in an environment that change rapidly and often.

The process of **endosporulation** (the formation of endospores) involves several stages. After the bacterial cell replicates its DNA, layers of peptidoglycan and protein are produced to surround the genetic material. Once fully formed, the endospore is released from the cell and may sit dormant for days, weeks, or years. When more favorable environmental conditions prevail, endospores **germinate** and return to active duty as vegetative cells.

Mature endospores are highly resistant to environmental conditions such as heat and chemicals and this permits survival of the bacterial species for very long periods. Endospores formed millions of years ago have been successfully brought back to life, simply by providing them with water and food.

Because the endospore coat is highly resistant to staining, a special method was developed to make them easier to see with a brightfield microscope. This method, called the **endospore stain**, uses either heat or long exposure time to entice the endospores to take up the primary stain, usually a water soluble dye such as malachite green since endospores are permeable to water. Following a decolorization step which removes the dye from the vegetative cells in the smear, the counterstain safranin is applied to provide color and contrast. When stained by this method, the endospores are green, and the vegetative cells stain pink, as shown in Figure 7.

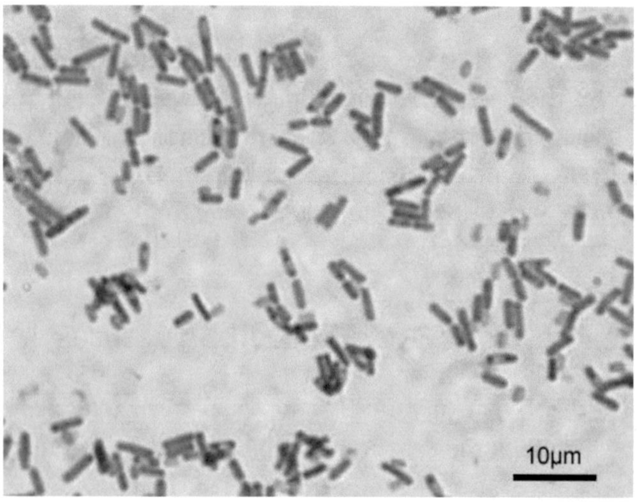

Figure 7. Bacterial cells with endospores, stained with the endospore stain.

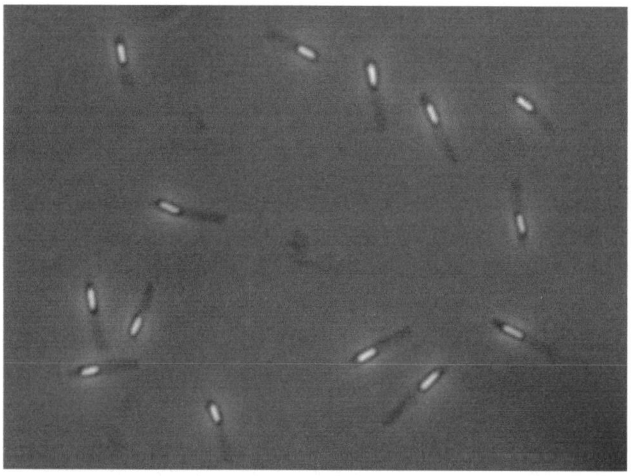

Figure 8. Bacilli with endospores viewed by phase-contrast microscopy.

Although endospores themselves are resistant to the Gram stain technique, bacterial cells captured in the process of creating these structures can be stained. In this case, the endospores are seen as clear oval or spherical areas within the stained cell. Endospores can also be directly observed in cells by using phase contrast microscopy, as shown in Figure 8.

Method

Because many differential staining methods require several steps and take a long time to complete, we will not be performing all of the differential staining methods discussed above.

Pre-stained slides will be used to visualize bacterial capsules, metachromatic granules, and acid-fast bacilli. Obtain one slide of each of the three bacteria listed in the table below. As you view these slides, make note of the "highlighted" structures. Your environmental isolate may have one or more of these cellular features, and learning to recognize them will aid in identification. These should all be viewed using the oil immersion objective lens.

Bacterium	Stain	Description or sketch of cells with the specified feature
Flavobacterium capsulatum	Capsule stain	
Corynebacterium diphtheriae	Methylene blue(metachromatic granules)	
Mycobacterium tuberculosis	Acid fast stain	

Gram Stain

All staining procedures should be done over a sink. The Gram stain procedure will be demonstrated, and an overview is provided in Table 1.

Table 1. Gram stain procedural steps.

Step	Procedure	Outcome
Primary stain(crystal violet)	Add several drops of crystal violet to the smear and allow it to sit for 1 minute. Rinse the slide with water.	Both Gram-positive and Gram-negative cells will be stained purple by the crystal violet dye.
Mordant (iodine)	Add several drops of iodine to the smear and allow it to sit for 1 minute. Rinse the slide with water.	Iodine "sets" the crystal violet, so both types of bacteria will remain purple.
Decolorization (ethanol)	Add drops of ethanol **one at a time** until the runoff is clear. Rinse the slide with water.	Gram-positive cells resist decolorization and remain purple. The dye is released from Gram-negative cells.
Counterstain(safranin)	Add several drops of safranin to the smear and allow it to sit for one minute. Rinse the slide with water and blot dry.	Gram-negative cells will be stained pink by the safranin. This dye has no effect on Gram-positive cells, which remain purple.

A volunteer from your lab bench should obtain cultures of the bacteria you will be using in this lab, as directed by your instructor. One of the cultures will be a Gram positive bacterium, and the other will be Gram negative. Below, write the names of the bacteria you will be using, along with the BSL for each culture:

____

Obtain two glass slides, and prepare a smear of each of the two bacterial cultures, one per slide, as demonstrated. Allow to COMPLETELY air dry and heat fix. Stain both smears using the Gram stain method. Observe the slides with a light microscope at 1,000X and record your observations in the table below.

Name of culture	Gram stain reaction	Cellular morphology	Arrangement

Gram Stain "Final Exam": prepare a smear that contains a mixture of the Gram-positive AND Gram-negative bacteria by adding a small amount of each bacterium to a single drop of water on a slide. Heat fix the smear and Gram stain it. You should be able to determine the Gram stain reaction, cellular morphology and arrangement of BOTH bacteria in this mixed smear. Your instructor may ask to see this slide and offer constructive commentary.

Endospore Stain

Only a few genera of bacteria produce endospores and nearly all of them are Gram-positive bacilli. Most notable are *Bacillus* and *Clostridium* species, which naturally live in soil and are common contaminants on surfaces. The growth of *Clostridium* spp. is typically limited to anaerobic environments; *Bacillus* spp. may grow aerobically and anaerobically. Endospore-forming bacteria are distinct from other groups of Gram positive bacilli and distinguishable by their endospores.

An overview of the endospore stain procedure is provided in Table 2.

Table 2. Endospore stain procedural steps.

Step	Procedure	Outcome
Primary stain(malachite green)	Add several drops of malachite green to the smear and allow it to sit for 10 minutes. If the stain starts to dry out, add additional drops.	Vegetative cells will immediately take up the primary stain. Endospores are resistant to staining but eventually take up the dye.
Decolorization(water)	Rinse the slide under a gentle stream of water for 10–15 seconds.	Once the endospores are stained, they remain green. A thorough rinse with water will decolorize the vegetative cells.
Counterstain(safranin)	Add several drops of safranin to the smear and allow it to sit for 1 minute. Rinse the slide and blot dry.	Decolorized vegetative cells take up the counterstain and appear pink; endospores are light green.

After staining, endospores typically appear as light green oval or spherical structures, which may be seen either within or outside of the vegetative cells, which appear pink.

The shape and location of the endospores inside the bacterial cells, along with whether the sporangium is either distending (D) or not distending (ND) the sides of the cell, are important characteristics that aid in differentiating among species (see Figure 9).

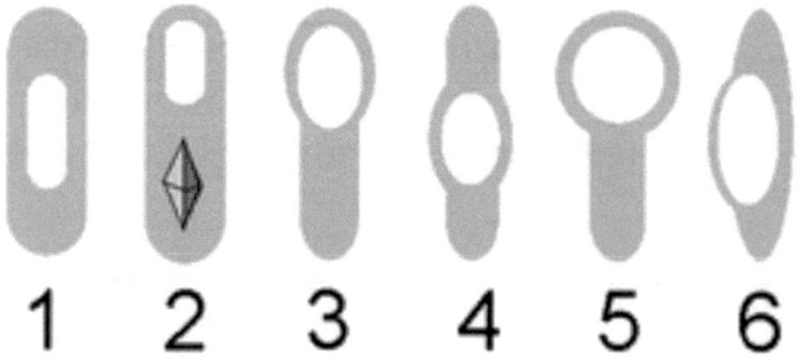

Figure 9

1. Oval, central, not distended (ND)
2. Oval, terminal, ND (and parasporal crystal)
3. Oval, terminal, distended (D)

4. Oval, central, D
5. Spherical, terminal, D
6. Oval, lateral, D

Endospores are quite resistant to most staining procedures; however, in a routinely stained smear, they may be visible as "outlines" with clear space within. If you observe "outlines" or what appear to be "ghosts" of cells in a Gram stained smear of a Gram-positive bacilli, then the endospore stain should also be performed to confirm the presence or absence of endospores.

A volunteer from your lab bench should obtain bacterial cultures for endospore staining, as directed by your instructor. Note that these will all be species of *Bacillus*. Prepare smears and stain each using the endospore staining technique. Observe the slides and note the shape and location of the endospore and the appearance of the sporangium (swollen or not swollen) in the table below:

Name of culture	Endospore Shape	Location	Sporangium

In addition, choose ONE of the cultures from above and Gram stain it. Record your results below in the spaces provided:

Name of Gram stained culture: ____

Gram stain reaction and cellular morphology: ________________________________

Are endospores visible in the Gram stained smear? ______________ If you see endospores, describe how they appear in the Gram stained preparation, and how this is similar to and different from what you see in the endospore stained preparation.

Metabolism, Physiology, and Growth Characteristics of Cocci

Metabolism

Like an animal or a plant, the life of bacteria involves a daily routine of thousands of chemical reactions, many devoted to the breakdown (catabolism) of substrates to extract energy or building materials. Other types of reactions utilize the energy and building blocks liberated during catabolism for synthesis reactions (anabolism). The term "metabolism" is an expression used to describe all of the chemical reactions that occur in a cell.

Bacteria rely on enzymes for their biochemistry, just as do other cell types. For bacteria, enzymes needed for metabolic reactions are either endoenzymes, which work within the cell, or exoenzymes, which are produced inside the cell and then transported to the outside where they facilitate the preliminary digestion of high molecular weight substrates that do not pass readily through the cell membrane.

All of this chemistry results in the production of biomolecules and waste, much of which is excreted by the cells into the surrounding environment. Detecting and identifying the biochemical products of metabolism provides us with a way to learn more about the physiologic and growth capabilities of bacteria, and also give us a way to differentiate among and/or identify species.

In *Bergey's Manual*, the definitive reference book on bacteria, determinations of identity or taxonomic group are based on many criteria. Gram stain reaction is usually the first criteria, followed by biochemical characteristics such as aerobic or anaerobic respiration, fermentation of various sugars, degradation of proteins and amino acids, and other cellular events. Intermediates or end products of these varied metabolic activities can be detected by performing biochemical assays on a bacterial culture. The results of these tests provide a biochemical profile, or "fingerprint," that can be used to classify or even identify the bacterial species. The outcomes of laboratory tests can also provide insight into physiology and what is needed to encourage and support bacterial growth.

One of the goals of Koch and Pasteur and their many associates was to develop methods to isolate and identify pathogens as the cause of a human disease. Although the limitations of this approach are now well known, the methods developed during the Golden Age period are still widely used in research and in clinical microbiology laboratories. From clinical specimens, isolated cultures are subjected to a battery of morphological and biochemical fingerprinting tests and compared to known outcomes. In this way, it is possible to identify potentially pathogenic bacteria and distinguish them from the usually helpful symbiotic microbiota.

Morphological Evaluation of Bacterial Isolates

The bacteria we will examine in this lab include species in different genera; *Staphylocccus, Micrococcus, Streptococcus, Enterococcus, and Neisseria.* At the cellular level, the one characteristic common to all of them is cellular morphology—all are cocci. They differ, however, in many other characteristics.

A volunteer from your lab bench should obtain cultures from your instructor, who will provide you with the species names. Write the name and BSL for each of the cultures below:

____

____

____

____

Colonial morphology

Although there are similarities, the bacteria we will examine in this lab have notable differences, starting with appearance of their colonies. For the bacteria listed in the table below, choose one species from each of the genera below to observe, and describe the colonial morphology of the bacteria in the table below. Note that up until about a decade ago, *Streptococcus* and *Enterococcus* were considered part of the same genus and are very similar with regard to both cellular and colonial morphology.

Bacterium	**Colonial Morphology**
Staphylococcus saprophyticus	
Micrococcus luteus	
Enterococcus faecalis	

Gram stain and cellular morphology

On a cellular level, all of the bacteria we will look at in this lab have a similar morphology, but there are significant differences in Gram stain reaction, cell size, and cellular arrangements. These differences help to target the particular genus of a bacterial sample.

Prepare smears of the three bacteria you examined (above) and Gram stain them. Also look at a prepared Gram stained slide of *Neisseria gonorrhoeae* (the causative agent of the STD gonorrhea) and describe what you see in the table below.

Bacterium(write in species name)Gram stain results(reaction, morphology, and arrangement)Staphylococcus ________________(also representative of other *Staphylococcus* spp.) Micrococcus luteus (also representative of other *Micrococcus* spp.) Enterococcus faecalis (representative of both *Entercoccus* and *Streptococcus* spp.) Neisseria gonorrhoeae(There is no culture of this – view the prepared slide)

Physiology and growth characteristics

Growing bacteria in culture requires consideration of their nutritional and physical needs. Food, provided in the media, is broken down by cells and used for energy and building biomass. Unlike eukaryotic cells, bacteria have options when it comes to making energy, which depend not only on the type of organic molecules in the food but also on the availability of oxygen as a final electron acceptor for respiration.

Respiration is the pathway in which organic molecules are sequentially oxidized to strip off electrons, which are then deposited with a final electron acceptor. Along the way, ATP is made. For many types of bacteria, oxygen serves as the final electron acceptor in respiration. Remarkably, oxygen is not always a requirement for respiration. For bacteria that live in environments with no air, alternative electron acceptors may take the place of oxygen.

Unlike the majority of eukaryotes, bacteria have options when it comes to making ATP. Aerobic respiration and anaerobic respiration generate ATP by chemiosmosis, and some bacteria may also ferment sugars, although the oxidation is not complete and energy is left behind. Chemical by-products and end products of these pathways are detectable and serve as the basis for many biochemical tests performed to identify bacteria.

Fermentation and anaerobic respiration are anaerobic processes—meaning that no oxygen is required for ATP production. Some bacteria have the capability (meaning they produce the appropriate enzymes) to use more than one, or even all three, of these pathways depending on growth conditions.

Based on whether oxygen is required for growth, bacteria can be considered to be either aerobes or anaerobes. However, because some bacteria may use more than one pathway, there are additional categories that describe a culture's requirement for oxygen in the atmosphere. The three major categories are:

> **Strict aerobe**—Bacteria that are strict aerobes must be grown in an environment with oxygen. Typically, these bacteria rely on aerobic respiration as their sole means of making ATP, but some may also ferment sugars.
>
> **Strict anaerobe**—These bacteria live only in environments lacking oxygen, using anaerobic respiration or fermentation to survive. For these types of cells, oxygen can be lethal because they lack normal cellular defenses against oxidative stress (enzymes that protect cells from oxygen free radicals).
>
> **Facultative anaerobe**—The most versatile survivalists there are. These bacteria typically have access to all three ATP-forming pathways, along with the requisite enzymes to protect cells from oxidative stress.

Additionally, overlapping categories include:

> **Microaerophile**—As the name implies, these bacteria prefer environments with oxygen, but at lower levels than normal atmospheric conditions. Often, microaerophiles also have a requirement for increased levels of carbon dioxide in the atmosphere and may also be called **capnophiles**. These bacteria make ATP by aerobic respiration and may also ferment sugars aerobically.

Aerotolerant anaerobe—These bacteria make ATP by anaerobic respiration and may also be fermentive. However, they are "tolerant" of oxygen because they may have cellular defenses against oxygen free radicals.

Tests that detect either components or end-products of these pathways may be used to assess a culture's overall oxygen requirement category. The following tests provide the information necessary to assess this growth characteristic.

The **catalase test** detects the ability of bacteria to produce an enzyme called catalase which is found in cells that live where there is air. Various chemical reactions in electron transport pathways create oxygen free radicals, which are electron-scavenging chemical species that can oxidize and potentially damage biomolecules in cells. One of these is hydrogen peroxide (H_2O_2), the substrate of the catalase enzyme which converts hydrogen peroxide to water and oxygen. The catalase test is performed by mixing a small amount of a bacterial culture with a drop of hydrogen peroxide on a slide. If the bacteria have the catalase enzyme, the substrate will be split, forming water and oxygen which is observed as bubbling when the gas is released (see Figure 1). A positive test result indicates that the bacteria live aerobically, and are likely to produce ATP by aerobic respiration. Strict aerobes, facultative anaerobes and microaerophiles may be positive for this test. Anaerobes (strict or aerotolerant) will be negative).

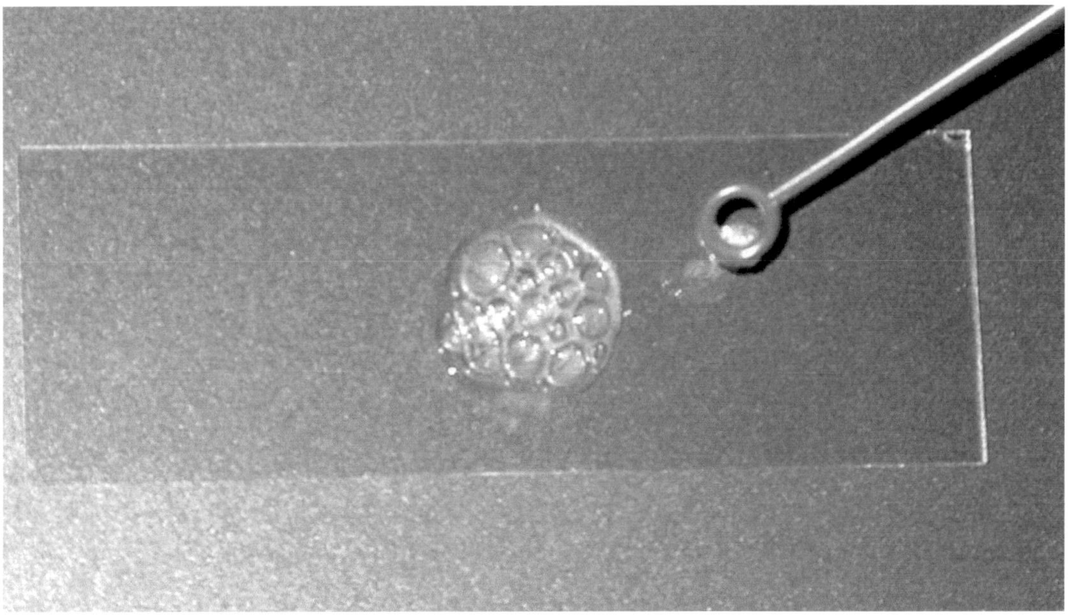

Figure 1. A positive catalase test, as evidenced by the release of "bubbles" of O_2.

The **oxidase test** identifies bacteria that produce cytochrome oxidase or indophenol oxidases, which are redox enzymes in the electron transport system that shuttle electrons to oxygen. The cytochrome system is usually only present in aerobic organisms that use oxygen as the final electron acceptor in respiration. There are several ways in which this test may be performed, but one of the simplest is to use a commercial test system, such as the BBL DrySlide Oxidase test, which consists of a card saturated with a chemical reagent that is colorless in its reduced state and turns dark blue when oxidized. The cytochrome oxidase enzymes donate electrons to the reagent, changing the color of the card from colorless to blue for a positive test (see Figure 2). Aerobic bacteria with a cytochrome-based electron transport system (similar to what is found in the mitochondria of eukaryotic cells) will be positive for this test.

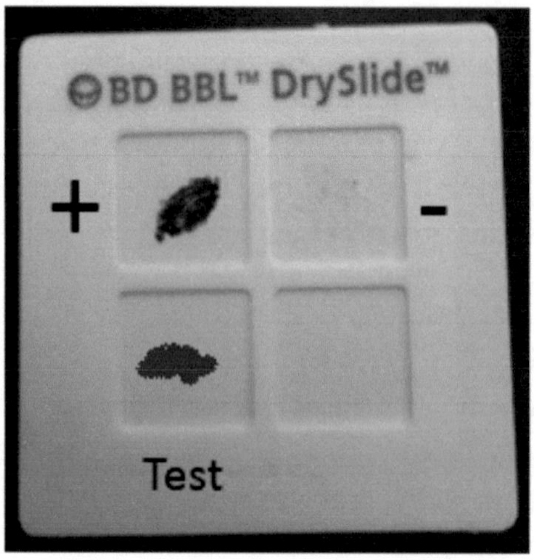

Figure 2. Positive and negative controls for the oxidase test. A positive test result is shown.

The **nitrate reduction test** detects reduced forms of nitrate, which occurs when bacteria use nitrate (NO_3) as a substitute for oxygen (O_2) during respiration. In the biogeochemical cycle known as the nitrogen cycle, nitrate reduction is the first step in a series of reactions collectively referred to as denitrification (Figure 3).

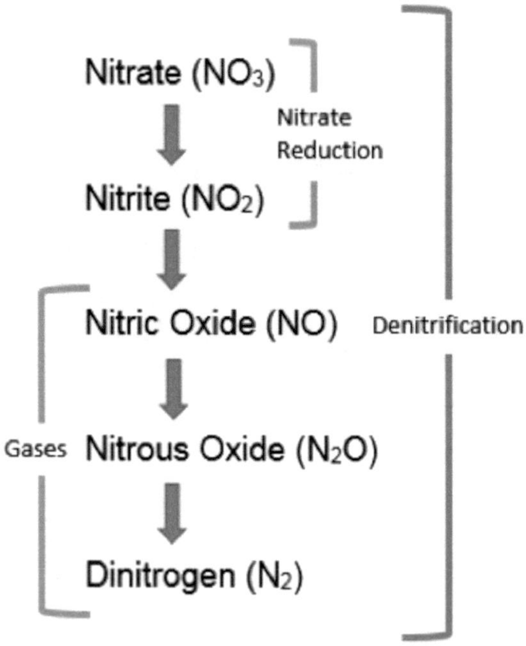

Figure 3. Sequential chemical reactions (reductions) occuring in the nitrogen cycle.

On an ecosystem scale, denitrification decreases the levels of NO_3 in soil and slows leaching of this substance into groundwater. On the other hand, denitrification may lead to an increase in N_2O, a "greenhouse gas" in the atmosphere and depletes nitrate from soil, which deprives plants and other microbes of this important nutrient. On a cellular scale, some bacteria reduce nitrate as a substitute for oxygen when they are in anoxic environments, and therefore, nitrate respiration can be a useful test for discriminating among bacterial species. This test is performed by subculturing bacteria to nitrate broth, a medium containing food and a source of nitrate available to serve as a final electron acceptor (as a substitute for oxygen for anaerobically respiring bacteria).

Nitrate reduction is demonstrated by adding chemicals that react with nitrite and noting development of a red color, which will occur if the bacteria reduced nitrate to nitrite. No color change after the chemicals are added might mean either the bacteria did not reduce the nitrate at all, or it may also mean the bacteria fully reduced the nitrate to N_2 (denitrification). This can be discriminated by adding zinc to cultures that do not change color when the reagents were added. Electrons donated by zinc will subsequently reduce any nitrate remaining in the broth to nitrite, and the broth will become red—therefore a negative test. If the bacteria already reduced all the nitrate to forms other than nitrite, no color change will occur, and this is considered a positive test. A positive nitrate reduction test is indication of an anaerobic lifestyle.

Triple Sugar Iron is a slant medium with two growth environments: aerobic (on the slant) and anaerobic (in the "butt"). The medium contains three sugars in varying concentrations and a pH indicator that turns yellow at pH measurements below 6.8, and a deeper red at pH measurements above 8.2. Bacteria that ferment typically produce one or more types of acid as a byproduct, therefore, fermentation (both aerobic on the slant and anaerobic in the butt) is noted as a change in the color of the media. The medium also identifies strict aerobes that only grow on the slant surface, and also bacteria that produce H_2S, either as a way to produce ATP anaerobically using sulfur or sulfate as a final electron acceptor, or as a result of the breakdown of proteins that contain high numbers of sulfur-containing amino acids (cysteine or methionine).

The results of this test are reported as appearance of the slant/appearance of the butt, using A to indicate acid reaction (yellow color), K to indicate an alkaline reaction, and NC to indicate no change in the medium. H_2S (detected as a blackening in the media) and the production of gas (CO_2) as a byproduct of fermentation are also reported if observed (see Figure 4).

As an example, and for practice, the interpretation and outcomes for the 4 TSI tests shown are provided in the table below. Note that many other possible reactions may also occur so proper interpretation of this test is important.

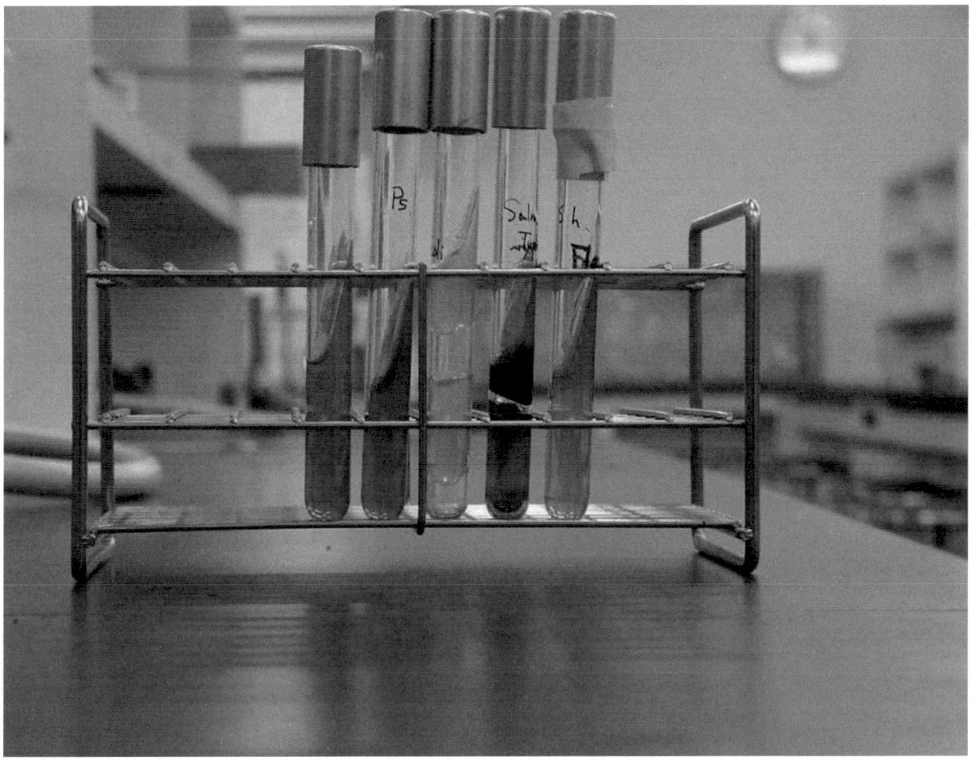

Figure 4. TSI outcomes produced by different bacterial cultures.

Table 1. TSI reactions shown in the cultures in Figure 4, from left to right.	
Outcome	Interpretation
Uninoculated control	For color comparison with inoculated samples
K/NC	Aerobic respiration (dark red on the slant) only. Bacteria are strict aerobes.
A/A; gas	Fermentation of all three sugars with CO_2 produced. Bacteria are facultative anaerobes.
K/A; H_2S	Aerobic respiration (dark red on slant), fermentation of glucose (acid only in butt), anaerobic respiration (black in butt). Bacteria are facultative anaerobes.
K/A	Aerobic respiration (dark red on the slant); fermentation of glucose (acid only in butt). Bacteria are facultative anaerobes.

How would you interpret the outcome of the TSI slant, the appearance of which is described below?

Appearance	Outcome and Interpretation
Slant is a dark red color, butt is yellow with noticeable cracking and bubbling.	

After inoculation and test procedures have been demonstrated, perform these tests on the bacteria listed in the table, and record the outcomes below:

Bacterium	**Catalase**	**Oxidase**	**Nitrate Reduction**	**TSI**
Staphylococcus (*aureus* **OR** *epidermidis*)				
Micrococcus luteus				
Enterococcus faecalis				

For each bacterium, determine if the test results provide evidence of aerobic or anaerobic respiration or fermentation, and indicate why you reached that conclusion. Then, based on your observations, state the logical growth category related to oxygen for each.

Staphylococcus ____________(write in species you tested)State what evidence from test results indicates the bacteria use this pathway. If there is no evidence, write "none."

Aerobic respiration	
Anaerobic respiration	
Fermentation	

Oxygen Growth Requirement Category______________________________________

Micrococcus luteus	State what evidence from test results indicates the bacteria use this pathway. If there is no evidence, write "none."
Aerobic respiration	
Anaerobic respiration	
Fermentation	

Oxygen Growth Requirement Category____

Enterococcus faecalis	State what evidence from test results indicates the bacteria use this pathway. If there is no evidence, write "none."
Aerobic respiration	
Anaerobic respiration	
Fermentation	

Oxygen Growth Requirement Category____

Differentiating Among Bacterial Species Based on Phenotypic Characteristics

All of the "volunteer" bacteria used for this experiment are in Bergey's Group 17 (Gram-positive cocci). From a medical perspective, some are considered primary or opportunistic pathogens, and others are nonpathogens. To distinguish among the cocci in this group, preliminary colonial and cellular characteristics along with growth patterns may be applied, as illustrated in the dichotomous key in Figure 5.

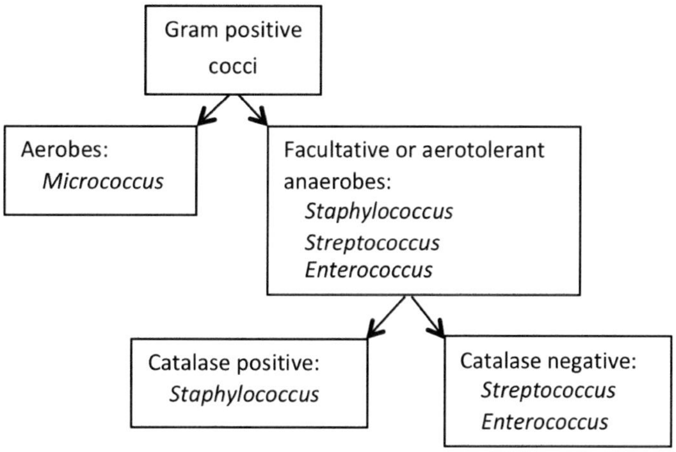

Figure 5. A dichotomous key to distinguish among Gram positive

Differentiating among *Staphylococcus* spp.

There are a large number of laboratory tests that facilitate differentiation among individual *Staphylococcus* spp. Cultures of three species of *Staphylococcus* have been provided. The following tests permit the differentiation of these species from one another. Note that an overnight incubation period is required for completion of these tests.

Coagulase test for differentiating *S. aureus* from other staphylococci

Staphylococcus aureus is known to cause several types of disease in human in addition to foodborne illness. Staphylococcal food poisoning may result from ingesting food contaminated with either the bacteria or a heat-stable enterotoxin produced by the bacteria.

S. aureus differs from most other species of staphylococci on the basis of its ability to produce the enzyme coagulase, which induces blood clot formation, along with other cell surface antigens such as Protein A. Bound coagulase is referred to as "**clumping factor.**" Coagulase and clumping factor can be identified using lab tests, and in a clinical laboratory this test is done routinely when Gram positive, catalase positive cocci are isolated from clinical specimens.

Because of the clinical significance, commercial kits are available to detect coagulase activity in bacterial cultures. One such kit is Staphaurex, a rapid test for the detection of clumping factor and protein A associated with *Staphylococcus aureus*. The kit includes a solution of white beads coated with fibrinogen and IgG, and special reaction cards that make the clumping of the beads obvious. When mixed with the reagent, coagulase-positive staphylococci induce the beads to rapidly form large clumps, which are easily seen against the black background of the card. The degree of clumping has to be interpreted by the observer, and this will be demonstrated in lab.

Mannitol Salt Agar: This is a selective medium for staphylococci and other **halotolerant** bacteria because the high concentration of salt (7.5%) inhibits the growth of bacteria susceptible to the effects of osmotic stress. In addition, the medium contains mannitol, which is a fermentable substrate for some bacteria, and phenol red as an indicator for acid. For those halotolerant bacteria that can grow on this medium, it is also possible to determine whether or not they ferment mannitol, by looking for a color change from red to yellow in the medium. The test is performed by streaking the bacteria over the surface of an MSA plate and incubating. Positive (left side) and negative (right side) results for mannitol fermentation are shown on the MSA plate in Figure 6.

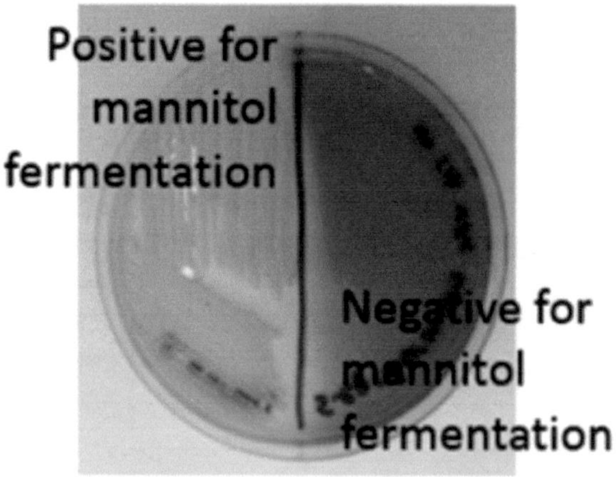

Figure 6. A Mannitol Salt Agar plate with cultures both positive and negative for mannitol fermentation.

Hemolysis: Some bacteria are known to produce enzymes that break down phospholipids and cause the cell membranes of red blood cells to rupture. Hemolytic bacteria then scavenge the hemoglobin released from the cell, typically to utilize the iron or other "growth factors" from inside the cell. Hemolysis can be observed by streaking bacteria across the surface of a Blood Agar Plate (BAP), which contains intact red blood cells. The BAP plate shown in Figure 7 is an illustration of β-hemolysis (beta hemolysis), seen as a clear area around the bacterial colonies.

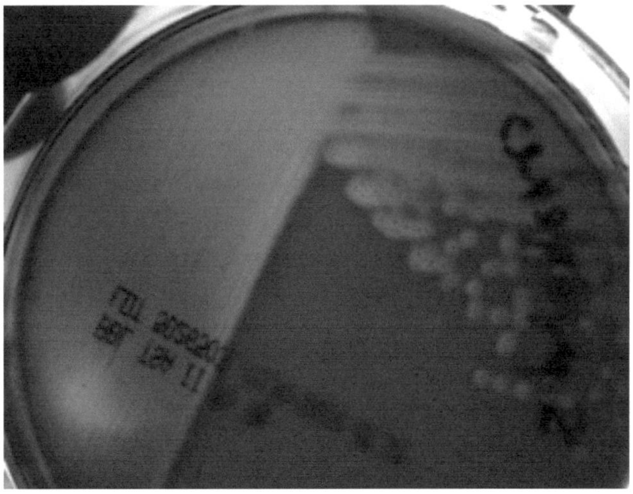

Figure 7. Beta-hemolysis on Blood Agar.

Figure 7 shows a BAP plate with colonies that are non-hemolytic. This is referred to as γ-hemolysis (gamma hemolysis). Not shown is a pattern of hemolysis called α-hemolysis (alpha hemolysis), which is not really a true lysis in that the red blood cell membrane is not ruptured, but merely "bruised." The hemoglobin, which mostly remains in the cell, is reduced to methemoglobin, which is a green color that can be seen surrounding the colonies growing on the BAP.

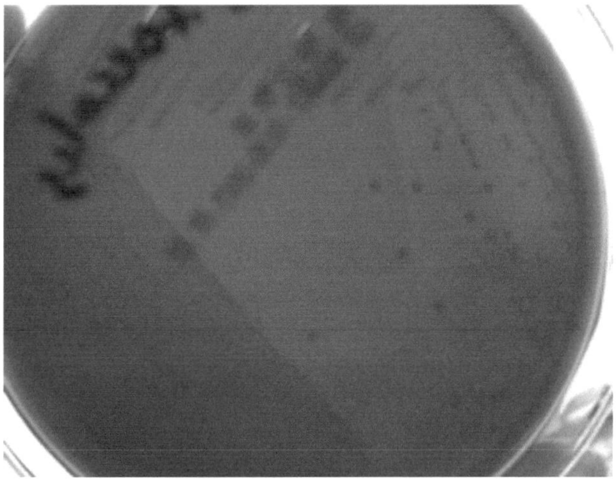

Figure 8. Gamma hemolysis (no hemolysis) on Blood Agar.

Urease: Many bacteria have the ability to hydrolyze urea, and some can do it more quickly than others. The enzyme urease is needed, which hydrolyzes urea to ammonia (a basic substance) and CO_2. Urease broth contains two buffers—urea, a tiny amount of food, and phenol red. This test is performed by inoculating bacteria into urease broth and incubating. If they produce urease rapidly, the urea in the broth is hydrolyzed and ammonia raises the pH of the broth. This process is detected by the pH indicator which turns deep pink, which is interpreted as a positive test. A positive (left tube) and a negative (right tube) urease test result are shown in Figure 9.

Figure 9. Positive and negative urease test results.

Obtain cultures of three species of *Staphylococcus*. Then, perform each of the tests above and record the outcomes in the table below, including both what the result looks like (color change, clear area around growth, etc.) and the interpretation of the outcome (positive, hemolysis, etc.). Note that these tests all require that the bacteria be incubated before the test can be completed.

Test	*S. aureus*	*S. epidermidis*	*S. saprophyticus*
Coagulase by Staphaurex			
Mannitol fermentation			
Hemolysis			
Urease			

From the results of the tests you performed on the three *Staphylococcus* spp., develop a dichotomous key (in the blank space below) that demonstrates how these tests can be applied to distinguish among the three species of *Staphylococcus* you tested:

Differentiating among streptococcal species

Chain-forming cocci (streptococci) are common members of the mammalian microbiota and sometimes cause disease. Many species inhabit the oral cavity and upper respiratory tract, while others are found in the GI tract. Philosophically, these two diverse habitats prompted taxonomists to split the GI dwelling streptococci into a separate genus, *Enterococcus*. Those species that inhabit the mouth remain in the genus *Streptococcus*. Streptococci in general are aerotolerant anaerobes, and can be distinguished from other types of Gram-positive cocci based on their negative response to the catalase test.

Bacteria in these two genera have many common characteristics. Because of their habitat, enterococci are tolerant to higher concentrations of bile. Bile is a yellow-green compound made up of bile acids, cholesterol, phospholipids, and the pigment biliverdin. It is produced in the liver, concentrated and stored in the gallbladder, and released into the duodenum after food is eaten, where it functions as a biological detergent that emulsifies and solubilizes lipids to help in fat digestion. The detergent action of bile also confers a potent antimicrobial activity. Thus, to survive in such an environment, the enterococci not only withstand the antimicrobial effects of bile, but also play a role in secondary bile metabolism in their host.

Bile Esculin Agar is a medium used to isolate and identify enterococci. This medium contains bile salts, which makes it selective for the bile tolerant enterococci, and esculin, which is an organic compound that some of the enterococci are able to chemically break down to glucose and esculetin. The latter substance combines with ferric ions in the medium, and form a complex which turns the both the colonies and the surrounding medium brownish black.

In a clinical laboratory, the medically important streptococci are identified by serological typing into Lancefield groups, which correlate to types of hemolysis observed when the bacteria are grown on Blood Agar Plates. The Group A streptococci, such as the pathogenic *Streptococcus pyogenes,* are β-hemolytic, while the Group D enterococci are typically non-hemolytic (γ-hemolysis).

Once the procedure has been demonstrated, subculture the two bacteria below to a Blood Agar Plate and Bile Esculin Agar plate and after incubation, observe the differences between them. Record the results below.

	Bile Esculin Agar	Hemolysis on Blood Agar
Streptococcus pyogenes		
Enterococcus faecalis		

On a separate sheet of paper below, **construct** a dichotomous key to show how the tests you performed may be used to distinguish among the different cocci we experimented with in this lab.

Table 2. Summary of Test Methods.

Test	Method
Catalase	Transfer bacteria to slide and add H_2O_2; observe for bubbles.
Oxidase	Smear bacteria to DrySlide oxidase card; watch for color change that occurs WITHIN 20 SECONDS.
Nitrate Reduction	Transfer 2 ml of Nitrate broth to a sterile culture tube, inoculate with bacteria and incubate. After incubation, add Nitrate A and B reagents – red is positive. IF NO CHANGE, add zinc. Red after zinc confirms negative result.
TSI	Stab butt and streak slant of TSI slant; incubate.
Coagulase	Place a drop of Staphaurex reagent in a circle on the test card. Add bacteria to the drop and mix; then rock the card.
Mannitol Salt Agar	Inoculate MSA plate and incubate.
Hemolysis on Blood Agar	Inoculate Blood Agar plate and incubate
Urease	Transfer 2 ml of Urease broth to a sterile culture tube; inoculate with bacteria and incubate.
Bile Esculin	Inoculate BE plate and incubate.

Metabolism, Physiology, and Growth Characteristics of Bacilli

The Diversity of Bacilli

Previously, we studied the colonial, cellular, and chemical characteristics of a group of bacteria that had a common cellular morphology (they were cocci), but were quite diverse metabolically. To continue the investigation, we will use a similar approach to examine the cellular and metabolic characteristics of a second group of bacteria, this time all bacilli.

Bacilli show great diversity at the cellular level, beginning with the overall size and shape of the cells themselves. Some bacilli are short and plump little rods, while others are extremely slender and long. Curved and spiral shapes are common. A few species of bacilli produce endospores or other types of cellular inclusion bodies, such as metachromatic granules and parasporal crystals.

Bergey's Manual initially divides the bacilli according to Gram stain reaction. Gram positive bacilli are further subdivided according to whether they form endospores, have filamentous growth or hyphae, and if they are acid-fast. Gram-negative bacilli are typically distinguished by size, shape, motility, and oxygen growth categories.

The group of bacteria included in this lab includes both Gram positive and Gram negative species. Using previous methods and additional tests, we will develop profiles of the bacilli as a way to distinguish among the various species.

Once again, although they are morphologically similar, the bacteria who have "volunteered" for this week's lab differ in many ways and can be distinguished from each other by observing morphological and physiological characteristics. And like last week, we will begin by observing colonial and cellular morphology.

Morphological Evaluation of Bacterial Isolates

The bacteria we will examine in this lab include both Gram positive and Gram negative species of bacilli, each in a different genus. Observable differences start at the cellular level.

A volunteer from your lab bench should obtain cultures from your instructor, who will provide you with the species names. Write the name and BSL for each of the cultures below:

____

____

____

____

Colonial morphology

Distinguishing among bacterial species begins with examining colonial morphology, which you will do using the cultures listed below:

Bacterium	**Colonial Morphology**
*Bacillus*__________________ (write in Species name)	
*Corynebacterium*____________ (write in Species name)	
Pseudomonas aeruginosa	

Gram stain and cellular morphology

Among the bacteria in this group, there are significant observable differences in cellular morphology that can be observed by staining the cells, beginning with the Gram stain. For each culture listed, Gram stain a smear you prepare, and record the Gram stain reaction, cellular morphology, and arrangement. Also if you observe any other cellular features, specifically endospores or inclusion bodies in the Gram stained smear, include those observations in your description as well.

Bacterium	**Gram stain results**
Bacillus __________________ (write in Species name)	
Corynebacterium ____________ (Write in Species name)	
Pseudomonas aeruginosa	

A major distinction among Gram positive bacilli is the production of endospores, which can be observed by endospore staining. For the two Gram positive species of bacteria, prepare a second smear and stain it using the endospore stain method. If endospores are observed, report the shape and location.

Bacterium	**Endospores present? (if yes, describe)**
*Bacillus*____________________ (write in Species name)	
*Corynebacterium*____________ (write in Species name)	

Physiology and growth characteristics

As previously done with the cultures of cocci, evaluate energy metabolism and physiological oxygen requirements for the bacteria in the table below by performing the catalase, oxidase, Triple Sugar Iron (TSI), and nitrate reduction tests.

Bacterium	Catalase	Oxidase	Nitrate Reduction	TSI
Bacillus ________________ (write in Species name)				
Pseudomonas aeruginosa				
*Salmonella*____________ ____ (write in Species name)				

From the outcome of the tests above, determine if there is evidence of aerobic or anaerobic respiration or fermentation, and indicate why you reached that conclusion.

Bacillus ______________	State what evidence from test results indicates the bacteria use this pathway. If there is no evidence, write "none."
Aerobic respiration	
Anaerobic respiration	
Fermentation	

Oxygen Growth Requirement Category______________________________________

Pseudomonas aeruginosa	State what evidence from test results indicates the bacteria use this pathway. If there is no evidence, write "none."
Aerobic respiration	
Anaerobic respiration	
Fermentation	

Oxygen Growth Requirement Category______________________________________

Salmonella ___________	State what evidence from test results indicates the bacteria use this pathway. If there is no evidence, write "none."
Aerobic respiration	
Anaerobic respiration	
Fermentation	

Oxygen Growth Requirement Category____

Differentiating Among Bacterial Species

In a clinical laboratory, differentiating the nonpathogenic microbiota found in the GI tract from pathogenic species has historically relied on phenotypic tests. The small and large intestines are home to the largest number of bacterial species, which contribute greatly to human health and disease. Generally, Gram positive cocci and bacilli are considered "probiotic" because of the health promoting benefits associated with these bacteria, while Gram negative species are commensals that can induce inflammation if their proliferation is not kept in check.

Many of the Gram negative bacteria residing in the GI tract are members of the Family *Enterobacteriaceae*. Bacteria in this family are facultative anaerobes that are usually oxidase-negative and ferment glucose, which distinguishes them from aerobic species of Gram negative rods. To differentiate among them, a set of tests with the acronym IMViC, which stands for: Indole, Methyl red, Voges-Proskauer, and Citrate (with the lower-case "I" thrown in to make it easier to say). The TSI and urease tests and others may also be performed at the same time to improve the identification process.

Examples of IMViC results for some of the *Enterobacteriaceae,* which illustrate how these tests can be used to differentiate among the bacteria in this group, are given in Table 1. As evidenced by the results shown for the two species of *Proteus*, metabolic differences may even exist among individual species within the same bacterial genus.

Table 1				
Species	Indole	Methyl Red	Voges-Proskauer	Citrate
Escherichiacoli	Positive	Positive	Negative	Negative
Shigella spp.	Negative	Positive	Negative	Negative
Klebsiella spp.	Negative	Negative	Positive	Positive
Proteusvulgaris	Positive	Positive	Negative	Negative
Proteus mirabilis	Negative	Positive	Negative	Positive

Sulfide–Indole–Motility (SIM)

This is another medium in which a number of different reactions may occur. In fact, this medium is used to determine if (1) the bacterium reduces sulfate and produces H_2S, which is evidence of anaerobic respiration; (2) the culture oxidizes the amino acid tryptophan and produces indole, and (3) the bacterium is motile. The medium contains sulfur (as the oxidizing agent) and ferrous ions (for H_2S

detection), tryptophan (for indole detection), and is prepared in tubes referred to as stabs or deeps (because they are not slanted). The medium also has a lower concentration of agar and is semi-solid, to detect bacterial motility. Bacteria are inoculated in this medium with an inoculating needle (as opposed to a loop) and stabbing the bacteria deep into the soft agar.

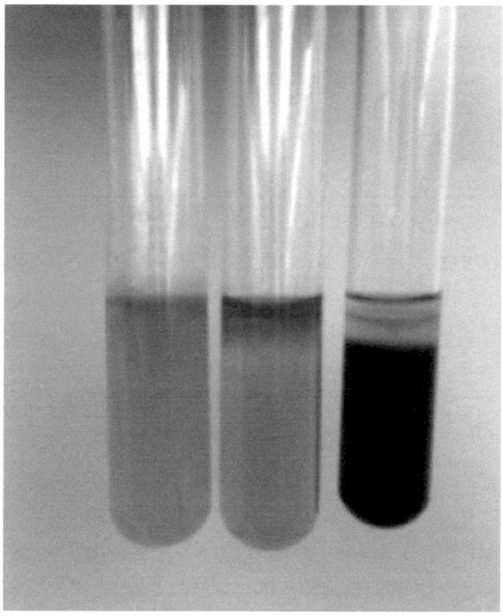

Figure 1. SIM medium inoculated with three different bacterial cultures.

After incubation, a positive test for sulfide turns the media black, just as for the TSI test. Tryptophan oxidation by the bacteria yields indole, which can be detected by adding a reactive chemical, which turns deep pink. Motility, which may be the hardest to determine, is noted by looking closely at the stab line. Motile bacteria swim through the semi-solid agar and the medium appears cloudy, obscuring the stab line. If the bacteria are not motile, the stab line is clearly visible. The outcomes for all three tests (sulfide, indole, and motility) should be noted. Figure 1 shows three SIM tubes, each showing different outcomes for the three tests. From left to right, the results are:

(1) Sulfide negative, Indole negative, Motility positive

(2) Sulfide negative, Indole positive, Motility positive

(3) Sulfide positive, Indole negative, and Motility positive

How would you interpret the results for the bacteria inoculated in the SIM medium shown below?

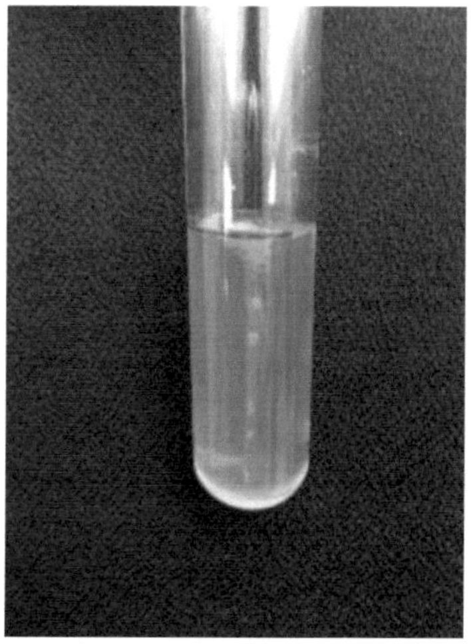

Sulfide: ______________________________

Indole: ______________________________

Motility: ______________________________

Methyl Red and Voges-Proskauer (MR-VP)

Fermentation of glucose by bacteria may produce one or more acids or alcohols. Probiotic lactic acid (homolactic) bacteria produce lactic acid as the major end product. Other fermentation pathways yield a combination of acids and/or alcohols. The organic end products of these pathways can be detected using chemical methods. The medium used for these tests, MR-VP broth is the same for both tests. The broth contains glucose as a fermentable substrate, and the end products are measured after incubation.

Methyl Red (MR)

Bacteria utilizing a mixed acid fermentation produce stable organic acid end products. After the bacteria grow, the pH indicator methyl red (which is red below pH 4) is added to the culture. If the broth is red after methyl red is added, the result is considered a positive test. For bacteria that produce alcohols and other less acidic metabolites, the indicator (and therefore the medium) turns orange to yellow, which is a negative result (see Figure 2a)

Voges-Proskauer (VP)

Some bacteria use a butanediol fermentation pathway when glucose is the fermented substrate. The end products of this type of fermentation include a variety of both acidic and non-acidic end products, including ethanol, butanediol, and acetoin. If acetoin is present, chemical reagents added to the broth will react with it and form a reddish-brown colored compound, which is considered a positive test result (see Figure 2b).

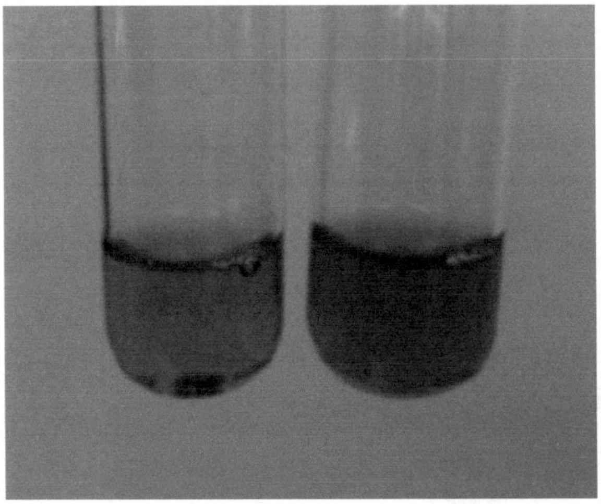

Figure 2a MR test

Figure 2b VP test

Citrate utilization

Some bacteria have an enzyme (citrate permease) that facilitates the transport of citric acid into the cell, and another enzyme, citrase, for citrate catabolism. These bacteria have the ability to grow on a medium (Simmons' citrate agar) that contains nothing more than citric acid, the first intermediate in the Krebs cycle of aerobic respiration, as a food source. Metabolism of citric acid releases carbon dioxide, which reacts with sodium and water in the medium to form a compound with a basic (alkaline) pH. In the presence of a base, the pH indicator in the medium (bromothymol blue) changes color from green to blue, which is a positive result for this test. The Simmons' citrate agar plate shown in Figure 3 illustrates both a positive (on the left) and negative (on the right) result for this test.

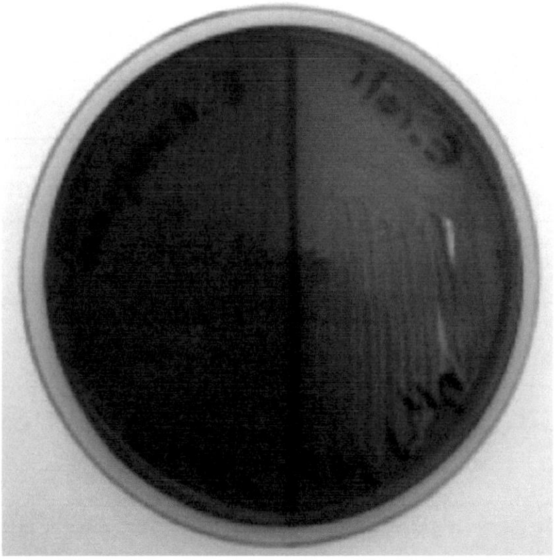

Figure 3. Citrate Utilization using Simmon's Citrate Agar.

Table 2. Summary of Test Methods	
Test	**Method**
SIM	With an inoculating loop, stab culture to the bottom of a SIM deep tube; incubate. **AFTER INCUBATION,** add 5 drops of indole (Kovac's) reagent and observe for deep red color.
Methyl Red (MR)	Transfer 2 ml of MR-VP broth to a sterile culture tube and inoculate with bacteria; incubate. **AFTER INCUBATION,** add 5 drops of methyl red reagent and observe for red color.
Voges–Proskauer (VP)	Transfer 2 ml of MR-VP broth to a sterile culture tube and inoculate with bacteria; incubate. **AFTER INCUBATION,** add 5-10 drops of Reagent A followed by 5-10 drops of Reagent B. Allow the tube to sit for at least 20 minutes and observe for red color.
Citrate	Inoculate the surface of a Simmons' citrate agar plate and incubate. Observe for blue color change.

After the tests have been demonstrated, perform them on the four *Enterobacteriaceae* listed in the table below.

Species	Indole	Methyl Red	Voges-Proskauer	Citrate
Escherichia coli				
Salmonella spp.				
Citrobacter freundeii				
Enterobacter aerogenes				

Construction of a dichotomous key to differentiate among this group of bacteria

In the space below, construct a dichotomous key to show how the tests you performed may be used to identify ALL of the different bacilli we used in this lab.

Microbiological Food Safety

Testing for Bacterial Contamination of Food

Bacteria are incredibly diverse and abundantly found in most of the natural world. The majority are beneficial to us in ways we may not fully realize or appreciate. A few, however, are not and will cause disease when we cross paths with them. Pathogenic (harm-causing) and potentially pathogenic bacteria may be found in unexpected places, such as in the food we eat, the water we drink or use for recreation, in soil, on surfaces in your home, and elsewhere.

Unfortunately for us, the things we eat and drink are fairly common vehicles for disease transmission. And, because food and drink pass through our digestive tract, the most common symptoms of a foodborne disease are abdominal discomfort or pain, nausea, diarrhea, and/or vomiting. Gastrointestinal illnesses caused by foodborne microbes range in severity from mild to extremely debilitating, even fatal. The biological agents responsible for this type of disease may be viruses, bacteria, fungi, protozoa, or helminthes.

To protect the public from disease, manufacturers and distributors of food consumed in the United States must prove that their food is pathogen-free before it can be offered for sale. Regulatory agencies at the local, state and federal levels (such as the Department of Agriculture and the Food and Drug Administration) require routine bacteriological testing to protect the public from acquiring a foodborne illness. Although many types of microbes may cause foodborne disease, the CDC and FDA currently considers the bacteria *Bacillus cereus*, *Campylobacter jejuni*, *Clostridium* spp., pathogenic strains of *Escherichia coli*, *Listeria monocytogenes*, *Salmonella* spp., *Shigella* spp., *Staphylococcus aureus*, *Vibrio* spp., and *Yersinia* spp.; the parasites *Cryptosporidium* and *Cyclospora;* and the Norwalk virus (norovirus) (http://www.cdc.gov/foodnet/index.html) to be the most common and of the greatest concern in the United States.

Although there are "rapid" methods available to detect bacterial contaminants in food that rely on DNA and antibody testing, plating samples on differential and selective culture media is a tried and true method. The disadvantage is that culture methods take more time, but the advantages include the simplicity of the tests and a higher level of both specificity and sensitivity.

The relatively low number of bacteria present in a food sample limits the sensitivity of all of the various types of tests available to evaluate food safety, including those based on culture. A preliminary step called enrichment culture may be used to amplify the number of bacterial pathogens, by pre-incubating the food sample in a non-selective medium that promotes growth of any bacteria that might be in the sample.

Many standard methods include a two-stage enrichment culture. The first step, or pre-enrichment, involves adding a specific amount (determined by weight, typically 10–25 grams) of the food to be tested in a large (100–250ml) volume of a non-selective broth medium. After an incubation period of 18–24 hours at 37 °C, a small sample of the enrichment culture is transferred to one or more types of selective media designed to inhibit growth of competing microbes while allowing the target pathogen to multiply. Many such formulations are also differential, in that growth of the bacterial target will cause a characteristic chemical change in the appearance of the medium, thus "differentiating" the pathogen from other possible contaminants, such as spoilage organisms, that might also be in the food.

We will be conducting our own investigation of food safety using a modified and scaled down adaptation of the standard laboratory methods, beginning with a pre-enrichment culture of food samples, followed by plated on several types of selective and differential media. Our determination of food contamination will be based on (a) growth of bacteria on the selective media and (b) observation of a specific biochemical reaction (usually a color change) characteristic for a particular type of pathogen. Note that these methods are based on bacterial phenotypes (traits), and more than one species of bacteria may have the same selective/differential traits. Therefore, definitive identification of a bacterium isolated from food requires additional testing.

Numerous media formulations are available that permit the isolation and identification of pathogenic bacteria in food. Using the media described in Table 1, we will be testing food for contamination with EHEC (enterohemorrhagic *E. coli*) and other strains of *E. coli, S. aureus, B. cereus, Salmonella*, and *Shigella*.

Table 1

Medium	Selective Agent(s)	Differential Agent(s)	Detection
MacConkey Agar (MAC)	Bile salts and crystal violet inhibits the growth of most Gram positive, non enteric bacteria.	Lactose and pH indicator	Gram negative enteric bacilli will grow; *E. coli* will produce pink colonies, *Salmonella* and *Shigella* spp. do not ferment lactose and colonies are colorless.
Sorbitol MacConkey Agar (SMAC)	Bile salts and crystal violet inhibits the growth of most Gram positive, non-enteric bacteria.	Sorbitol and pH indicator	Gram negative enteric bacilli will grow; *E. coli* 0157:H7 does not ferment sorbitol and colonies are colorless
Mannitol Salt Agar (MSA	7.5% sodium chloride inhibits the growth of most Gram negative bacteria	Mannitol and pH indicator	Salt tolerant bacteria grow; *S. aureus* ferments mannitol and colonies are yellow; *B. cereus* does not ferment mannitol and colonies are deep red.

Pre-enrichment to Promote Bacterial Growth

Note: this will require time in addition to your regular lab period to complete. Because both the presence and type of bacteria that may be in the food is unknown, BSL2 containment practices should be used throughout the entire procedure.

Samples of foods will be available in the laboratory up to a week before your scheduled lab period.

Transfer 5 ml of Tryptic Soy Broth to a disposable plastic culture tube. You should select one food sample for testing. Prepare an enrichment culture of the selected food item, by transferring a small amount of the food to the broth in the culture tube, using aseptic technique. Place a cap on the tube, mix the contents fully, and place in the incubator at 37°C.

A minimum of 18 hours after starting the enrichment culture (one day after the enrichment culture is started is preferred), and **no later** than the day before the scheduled period for this investigation, return to the lab, and use an inoculating loop to subculture samples from the enrichment culture to each of the three types of selective and differential media described in Table 1. Use the streak plate method for all of the plates, so isolated colonies will form. Appropriately dispose of the enrichment culture as a potential biohazard.

Return the streak plates to the incubator for observation and further investigation during the lab period.

Food sample tested ____

Cooked or raw? ____

How was this food item stored before testing? ____________________________

Medium	Growth on medium? (yes or no)	Appearance of colonies/medium if growth occurred
MacConkey Agar (MAC)		
Sorbitol MacConkey Agar (SMAC)		
Mannitol Salt Agar (MSA)		

For each of the selective and differential media on which bacterial colonies are observed, indicate what the appearance of the colonies indicates in terms of the possible type(s) of bacteria present in the food sample.

Medium	Observations that indicate contamination of food sample with potential food borne pathogens
MacConkey Agar (MAC)	
Sorbitol MacConkey Agar (SMAC)	
Mannitol Salt Agar (MSA)	

The growth and appearance of colonies on selective and differential media is an indicator of the presence of specific bacterial pathogens, but these results must be confirmed before reporting that food is contaminated and ingestion may initiate a foodborne disease. Therefore, perform additional tests to confirm that the colonies observed on the selective media are potentially pathogenic bacteria. These tests include:

Gram stain of representative colonies (all)

Coagulase test to confirm *Staphylococcus aureus* (*S. aureus* is coagulase +)

Colonial morphology and endospore stain to confirm *Bacillus cereus*

TSI test to confirm *E. coli, Salmonella*, and *Shigella*

Expected and Experimental Results

For each bacterium, indicate the EXPECTED outcome of the Gram and Endospore stains and TSI tests which would confirm the identity of the potential foodborne pathogen.

Expected Gram stain reactions:

E. coli ____________________ *Salmonella spp.* ____________________

S. aureus ____________________ *Shigella* spp. ____________________

B. cereus ____________________

Expected Endospore stain reactions:

B. cereus ____

Expected TSI reactions:

E. coli ________________ *Salmonella spp.* ________________ *Shigella* spp. ________________

For each type of colony found growing on the various types of selective and differential media, perform the appropriate confirmatory (or indicatory) tests, and record those results in the results table.

Results of Confirmatory Tests

Potential Pathogen	Found on which medium?	Additional test(s)	Outcome of test(s)	Present in Food Sample?
E. coli				
E. coli 0157:H7				
S. aureus				
B. cereus				
Salmonella spp.				
Shigella spp.				

Once you know whether the pathogens we tested for in the lab were present in the food sample you tested, compare your results with those of others in your lab who tested other types of food.

Approximately how many of the food samples that were tested by your lab group were contaminated with one or more of these bacteria? List the foods with contaminants below:

Reflect on the significance of the outcome of this investigation, and write your thoughts below. Before the class ends you may be asked to share your reflection as part of a larger discussion on what it means to find bacteria (pathogenic or otherwise) in food you might eat.

The War on Germs

Kill Them Before They Kill You!

Since the Golden Age of microbiology, when the connections between bacteria and disease were first revealed and Semmelweis started washing his hands, Pasteur postulated the Germ Theory, and Lister promoted aseptic surgery, we have become obsessed with destroying microbes. As a result, companies that make products to "kill" microbes—whether to have a cleaner home or to cure bacterial infections—make a killing off of us as we rush to acknowledge the effectiveness of marketing campaigns that tell tales about "evil germs" that must be "killed on contact."

In reality, promoting the need for humans to live in a germ-free world is misguided and in some cases perhaps even dangerous. The overwhelmingly beneficial nature of our microbiome and its metagenome is now well established. Overuse of personal and environmental hygiene products and antibiotics has resulted in the logical evolution of bacterial resistance to them. As Friedrich Nietzsche (and Kelly Clarkson) famously said, "That which does not kill us makes us stronger." This certainly bears true for bacteria. Some bacterial pathogens are now resistant to every type of antibiotic available. Bacteria acquire antibiotic resistance through mutation and natural selection, and once such traits are acquired, they are shared with other bacteria by horizontal gene transfer. Infectious diseases once thought defeated, such as tuberculosis and pneumonia, are again climbing the ranks on the mortality charts.

We are destined to live in a microbial world. Prudent use of antimicrobial products is a necessity, particularly in healthcare settings where sanitation practices and infection control are essential. Available are a large number of chemical and physical agents that are used for **sterilization** (killing or removal of all microorganisms) and **disinfection** (reducing the numbers of microorganisms) that are widely used to control microbial growth. But how do you know if they work?

Chemicals used to kill microbes damage cell components through chemical reactions with proteins, membranes, or other parts of bacterial cells. Heat, cold, and radiation are physical agents that inhibit or inactivate microbes in food, on surfaces, and even in the air. Bacteria exhibit a wide range of susceptibilities to these agents. For example, Gram-positive bacteria can withstand higher heat and more radiation, while Gram-negative bacteria tend to be more resistant to chemical destruction. Microbial susceptibility to antibiotics varies widely.

Selective removal of pathogenic microbes that spares or minimally damages body cells has long been the goal of medicine. Paul Ehrlich was searching for just such a "magic bullet" when he found the first real antibiotic, the arsenic-based compound Salvarsan for treatment of syphilis, in 1910. Now we realize there can be no magic bullet, because to one extent or another, antibiotics wreak havoc on our bacterial symbionts as much as the targeted pathogen, and those bacteria are as much a part of us as are our own cells.

Some antibiotics are effective against a **narrow spectrum** (or range) of bacteria, such as the antibiotic isoniazid which is only used to treat infections caused by mycobacteria. **Broad spectrum** antibiotics work against a wide range of taxonomic groups. In medicine, the preferred approach is to minimize damage to the microbiota by applying a narrow spectrum drug first, which also decreases the risk of promoting antibiotic resistance. Testing pathogen susceptibility to several types of antibiotics while the bacteria grow in culture is the preferred way to evaluate which to use for chemotherapy of the infection.

Several culture-based methods have been developed for this purpose. One of the most widely used methods is the disk diffusion test, used to assess the antimicrobial activity of chemical agents. The Kirby-Bauer assay, used in clinical laboratories to evaluate which antibiotics are effective against a bacterial pathogen, is a standardized form of disk diffusion test that follows a strict protocol (e.g. type of medium used, temperature and time of incubation) so the results can be compared across labs.

Kirby-Bauer Disk Diffusion Assay for Antimicrobial Susceptibility

After the widespread use of antibiotics began in the 1950s, one of the first observations made was that they didn't work for every infection. Thus test methods to evaluate susceptibility were developed. The earliest versions were based on broth dilution methods, which although very useful for determining the level of susceptibility to various concentrations of the drug, were time and labor-intensive to perform. And so the disk diffusion method was developed.

The disk diffusion method was widely adopted as the standard method for antimicrobial testing by clinical labs in the US by the end of the 1950s. Basic procedures were modified to suit the locally available resources and expertise, and interpretation of results began to vary widely by lab.

Kirby and colleagues (including Bauer) were the first to propose that standardization of the method was necessary for uniformity across labs, and developed the method that still bears their names: the Kirby-Bauer assay.

Of major importance to the success of this method is to inoculate test plates with approximately the same number of bacterial cells each time the test is performed. A relatively simple way to achieve a uniform number of cells in the inoculum is to first prepare suspensions of bacterial cells in sterile saline, which may then be compared, either using an instrument such as a spectrophotometer or by direct visual comparison, to an optical standard matching a known concentration of bacteria. Typically a 0.5 McFarland standard is used. This is equivalent to a bacterial suspension containing between 1×10^8 and 2×10^8 CFU/ml of *E. coli*.

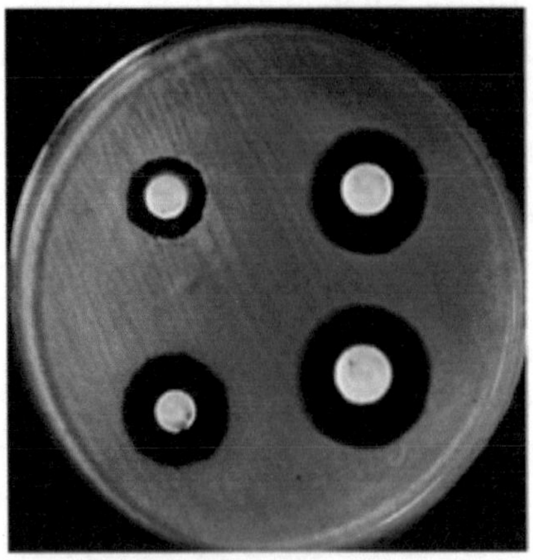

Fig. 1- PETRIDISH -1 (S. aureus treated with Pet. Ether and chloroform extract of *S. grandiflora*)

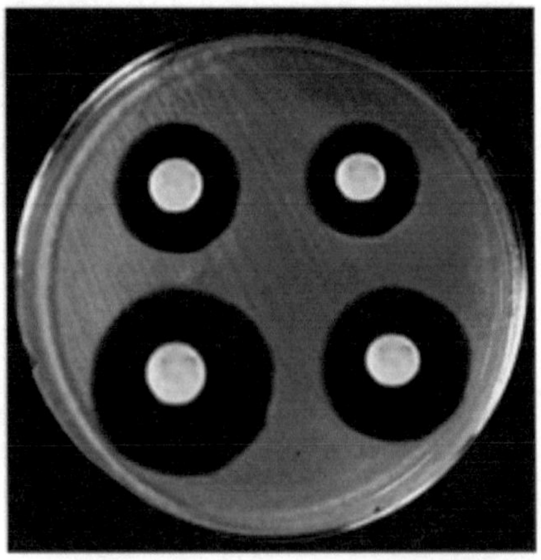

Fig. 2- PETRIDISH -2 (S. aureus treated withAcetone and Ethanol extract of *S. grandiflora*)

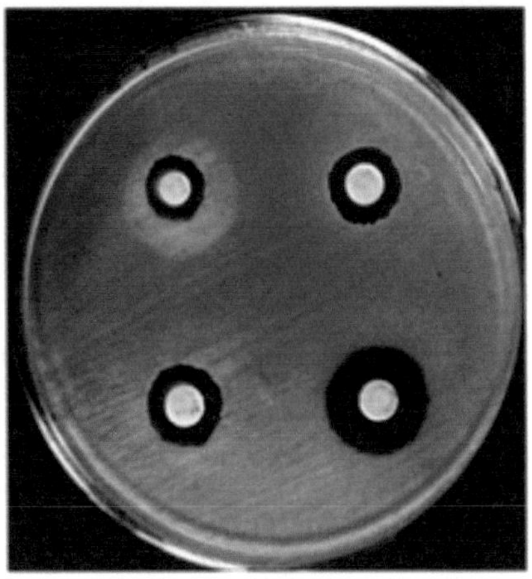

Fig. 3- PETRIDISH -3 (E.coli treated with Pet. Ether and Chloroform extract of *S. grandiflora*)

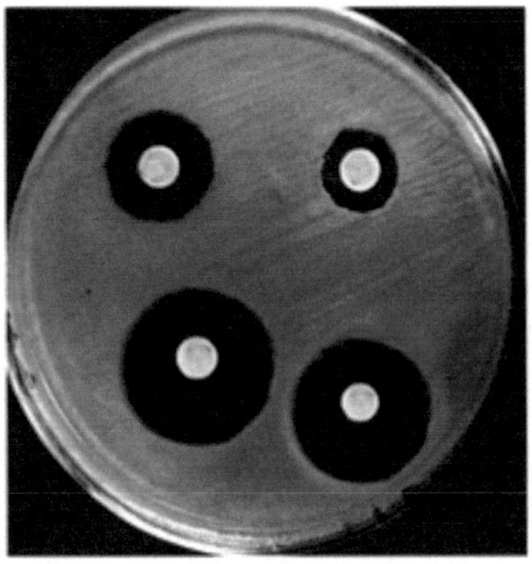

Fig. 4- PETRIDISH -4 (E.coli treated withAcetone and Ethanol extract of *S. grandiflora*)

Figure 1. Zone of inhibition in a bacterial lawn.

Standardized cell suspensions are then inoculated onto the surface of a specific type of medium using a sterile swab. The plates are inoculated to completely cover the surface of the medium with bacterial cells, which then grow into a lush "lawn" of colonies. Once the lawn is "seeded," a filter paper disk saturated with the antibiotic is placed on the lawn. The antibiotic diffuses into the medium at the same time that the bacteria are trying to grow. If the bacteria are susceptible, an area of clearing called the "zone of inhibition" will be seen in the lawn surrounding the disks at the end of the incubation period. The definition of susceptibility and resistance varies with the type of test—either the bacteria are observed to grow right up to the edge of the antibiotic disk, or a small zone of inhibition that falls within a measured size range is seen. The agar plate in Figure 1 shows variations in the size of the zones of inhibition for four antibacterial chemicals. For the Kirby-Bauer assay, the size of each zone must be carefully measured.

As demonstrated, prepare suspensions in sterile saline to match a 0.5 McFarland standard using the bacteria listed below. For each, also indicate if they are Gram-positive or Gram-negative.

Escherichia coli ______________________________

Staphylococcus aureus ______________________________

Pseudomonas aeruginosa ______________________________

Once made, the three standardized suspensions of these bacteria will be used again later in this lab, so do not immediately discard them.

For uniformity across labs, the same type of medium must be used by each lab. The standard method calls for Mueller-Hinton agar (MHA) plates, which is a special formulation prepared with a lower concentration of agar at a specific pH to facilitate diffusion of the antibiotics into the medium.

Obtain three MHA plates, and as demonstrated, inoculate each with one of the bacteria listed above, using a sterile swab soaked in the corresponding bacterial suspension.

Using forceps sterilized in alcohol, place one disk of each antibiotic to be tested on all three of the inoculated plates, far enough away from each other so that the potential zones of inhibition will not overlap.

The standard method calls for the plates to be incubated at a temperature range of 35°C ± 2°C for 18–24 hours. This may be modified to accommodate your laboratory schedule, as directed by your instructor. Make a note of the time and temperature for incubation if it is a modification of the standard method, below:

Incubation time and temperature: ______________________________

The four antibiotics we will be testing have different cellular targets, and therefore bacterial responses to them will vary. The bacterial cell wall is often a target of disruption for antibiotics. Two of the bacteria above have the same Gram stain reaction, but *Pseudomonas* spp. generally have a broader range of resistance than other types of bacteria. Research, and then below, list two strategies employed by pseudomonad bacteria to resist destruction by antibiotics:

____

____

____

____

____

Complete the table below with the mechanism of action for each of the antibiotics we'll be testing. Also, use a website such as Drugs.com or Rxlist.com to research the medical use (indications for use) for the antibiotic, and include that information in the table.

Antibiotic	Bacterial cell target/ Mechanism of action	Indications for use	Contraindications for use (including side effects)
Cefixime (Suprax)			
Tetracycline (Panmycin)			
Azithromycin (Zithromax)			
Ciprofloxacin (Cipro)			

After incubation, look for zones of inhibition around each disk in all three lawns. If a zone is observed, measure the diameter in mm using a metric ruler and record the measurement in the table below. Using the **table of standard values** (Table 1), compare the measured size of the ZOI and determine if the zone size indicates the bacteria are resistant, susceptible, or have an intermediate susceptibility to each of the antibiotics.

	E. coli		*S. aureus*		*P. aeruginosa*	
Antibiotic	ZOI in mm	Result	ZOI in mm	Result	ZOI in mm	Result
Cefixime						
Tetracycline						
Azithromycin						
Ciprofloxacin						

Based on what you learned about each antibiotic in terms of its medical uses and patterns of susceptibility, consider which of the four antibiotics tests you would prescribe for the following types of infections, if you were the medical professional in charge of the case:

A urinary tract infection (UTI) caused by *E. coli* ________________________________

A wound infection caused by *S. aureus* ________________________________

A respiratory tract infection caused by *P. aeruginosa* ________________________________

Using the Disk Diffusion Method to Test Chemicals for Antibacterial Action

The basic principles of the Kirby-Bauer method can also be applied to investigate the antimicrobial properties of solutions with known or suspected actions against bacteria.

Disinfectants and **antiseptics** are chemicals used for microbial control in many settings. As with antibiotics, microorganisms differ substantially in their susceptibility to the chemical agents we use to reduce the number of microbes on inanimate surfaces (disinfectants) or living tissues (antiseptics). To evaluate their potency, effectiveness is compared to that of a standard disinfectant such as phenol, which was the chemical first used and endorsed by Joseph Lister for aseptic surgery. This type of comparison can be done using broth dilution methods, but may also be accomplished by disk diffusion assay, in which the effectiveness of a particular chemical agent is assessed in a direct comparison of the size of the zone of inhibition to a positive control like phenol, which has a known and powerful antibacterial effect.

Obtain three MHA plates and label each plate with the name of one of the test bacteria (*E. coli, S. aureus,* and *P. aeruginosa*). Using the 0.5 McFarland-adjusted cell suspensions made previously, prepare lawns of the three bacteria on the surface of the MHA plates.

Available in lab will be several samples of chemicals with known or suspected antimicrobial properties. Choose three that you would like to investigate, and list your choices below. If the agent is a commercial product, also list the "active ingredient" from the label of the product. Also, record the positive control you will be comparing your selections against, according to your instructor.

Chemical Agent	Active Ingredient (if known)

Positive Control	Active Ingredient

On the bottom of the Petri dish holding the bacterial lawns, use a marker to draw intersecting lines that divide the plate into four sections. Label one section (+) to represent the positive control. Label the three other sections with the name or abbreviation of the test chemicals you selected.

Using forceps sterilized in alcohol, remove a sterile filter paper blank disk from its container and soak it in the positive control solution. Remove it from the solution and allow the excess liquid to drip back into the container. Place the soaked disk in the center of the appropriately labeled section on one of the plates, and then repeat the process for the other two plates.

Once the positive control disks have been placed, soak blank disks in the test solutions you chose and place them on the three lawns. Use the same three test solutions on each of the lawns.

Incubate the plates (overnight at 37°C is preferred). After the incubation period, examine the disks in the lawns for a zone of inhibition. Measure the diameter of the ZOI for the positive control first, as this is the standard by which you will be assessing the effectiveness of the test solutions. Record your results in the table below.

Solutions tested (write name)	*E. coli*	*S. aureus*	*Ps. aeruginosa*
Pos. control ________________			
Test 1 ________________			
Test 2 ________________			
Test 3 ________________			

List the chemical agents you tested that were at least as effective against each bacterium as the positive control:

Bacteria Effective test chemicals

E. coli ______________________________

S. aureus ______________________________

P. aeruginosa ______________________________

Of the chemical agents you tested, which one shows the greatest overall range of antibacterial action?

____

____

Which bacteria showed the greatest degree of resistance to the chemical agents tested?

____

Were any of the results a "surprise" to you (for example, perhaps something you thought would be a good disinfectant was not, or one of the natural products tested was more effective than predicted?). Reflect below, and you may be asked to share your thoughts on this during a discussion of the results.

____

____

____

Using Antibiotic Susceptibility for Bacterial ID

Browsing through *Bergey's Manual* reveals that susceptibility or resistance to various antibiotics can be useful when it comes to discriminating between bacterial species. Two examples are the novobiocin resistance test for *Staphylococcus* spp. and the "Taxo A" or the bacitracin susceptibility test for *Streptococcus* spp.

Novobiocin resistance test

This test was originally developed as a test to identify *S. saprophyticus*, which is a coagulase-negative staphylococci and the second most common cause of urinary tract infections (after *E. coli*). After the procedure has been demonstrated, perform the test using the cultures of *Staphylococcus aureus* and *Staphylococcus saprophyticus* provided.

Prepare standardized suspensions of *S. aureus* and *S. saprophyticus*, by transferring 2 ml of sterile saline to a culture tube, then adding colonies until the turbidity (cloudiness) in the solution matches that of a 0.5 McFarland optical standard.

Obtain one Mueller-Hinton agar (MHA) plate, and on the bottom of the Petri dish, draw a line with a marker to divide the plate in half. Label each half with the name of one of the two staphylococci.

Soak a sterile swab in one of the bacterial suspensions, and transfer it to the appropriately labeled side of the divided plate. Spread the suspension to create a confluent layer of bacterial cells which will grow into a lawn of bacterial colonies. Repeat the process with the other bacterial suspension on the other half of the plate.

With forceps sterilized in alcohol, place one novobiocin disk (5 μg) in each of the two lawns. Incubate the plate for a minimum of 24 hours. After incubation, examine the lawn around each disk. If a zone of inhibition is noted, measure the diameter of the zone with a metric ruler, in mm. For this test, bacteria are considered **positive** for **novobiocin resistance** if there is no zone of inhibition, or if the zone size is less than 12 mm. Susceptibility is a negative test. Record your results below.

Bacteria	ZOI diameter (mm)	Susceptible or Resistant?	Positive or Negative outcome
S. aureus			
S. saprophyticus			

Table 1. *Table of Standard Values: Kirby-Bauer Disk Diffusion Test*

Antibiotic	Spectrum	Disk	Zone diameter nearest whole mm		
			Resistant	Intermediate	Susceptible
β-**LACTAMS** (penicillins)					
Carbenicillin	Pseudomonas	100 μg	<13	14–16	>16
	Gram negatives	100 μg	<19	20–22	>23
Methicillin	Staphylococci	5 μg	<9	10–13	>14
Mezlocillin	Pseudomonas	75 μg	<15		>16
	Other Gram negs	75 μg	<17	18–20	>21
Penicillin	Staphylococci	10 units	<28		>29
	Enterococci	10 units	<14		>15
Piperacillin	Pseudomonas	100 μg	<17		>18
	Other Gram negs	100 μg	<17	18–20	>21
β-LACTAM/ β-LACTAMASE INHIBITOR COMBINATIONS					
Amoxycillin/ clavulanate	Staphylococci	20/10 μg	<19		>20
	Other organisms	20/10 μg	<13	14–17	>18
Piperacillin/ tazobactam	Pseudomonas	100/10 μg	<17		>18
	Other Gram negs	100/10 μg	<17	18–20	>21
	Staphylococci	100/10 μg	<17		>18
CEPHALOSPORINS					
Cefotaxime		30 μg	<14	15–22	>23
Cefixime		30 μg	<15	16–18	>19
Ceftriaxone		30 μg	<13	14–20	>21
Cefuroxime oral		30 μg	<14	15–22	>23
CARBAPENEMS					
Imipenem		10 μg	<13	14–15	>16
MONOBACTAMS					
Aztreonam		30 μg	<15	16–21	>22
GLYCOPEPTIDES					
Vancomycin	Enterococci	30 μg	<14	15–16	>17
	Other Gram pos	30 μg	<9	10–11	>12
AMINOGLYCOSIDES					
Gentamicin		10 μg	<12	13–14	>15
Streptomycin		10 μg	<11	12–14	>15
Tobramycin		10 μg	<12	13–14	>15
MACROLIDES					
Azithromycin		15 μg	<13	14–17	>18
Clarithromycin		15 μg	<13	14–17	>18
Erythromycin		15 μg	<13	14–22	>23
TETRACYCLINES					
Doxycycline		30 μg	<12	13–15	>16
Minocycline		30 μg	<14	15–18	>19
Tetracycline		30 μg	<14	15–18	>19

QUINOLONES					
Ciprofloxacin		5 µg	<15	16-20	>21
Nalidixic acid		30 µg	<13	14-18	>19
Norfloxacin		10 µg	<12	13-15	>16
Ofloxacin		5 µg	<12	13-15	>16
OTHERS					
Chloramphenicol		30 µg	<12	13-17	>18
Clindamycin		2 µg	<14	15-20	>21
Nitrofurantoin		300 µg	<14	15-16	>17
Rifampin		5 µg	<16	17-19	>20
Sulfonamides		250/300 µg	<12	13-16	>17
Trimethoprim		5 µg	<10	11-15	>16
Trimethoprim/ sulfamethoxazole		1.25/ 23.75 µg	<10	11-15	>16

Note Adapted from: Clinical Laboratory Standards Institute. (2006). Performance standards for antimicrobial disk susceptibility tests; Approved standard 9ed. CLSI document M2-A9. 26:1. Clinical Laboratory Standards Institute, Wayne, PA.

Epidemiology and Public Health

Epidemiology is a science that studies the causes and effects of health-related events as they occur in populations. Disease, defined as a deviation from health, is one such health-related event of concern to epidemiologists, so in that regard, epidemiology is often thought of as the study of disease in populations.

Although the historical origins of epidemiology as a science are investigations of epidemics of infectious disease, modern epidemiology has expanded to not only include contagious diseases, but also environmental connections to disease states and even accidental injuries. Epidemiologists gather data on the frequency of various diseases in populations, and correlate risk factors associated with disease development.

The information compiled by epidemiologists provides the foundation for the concept of "public health." The focus of public health is to prevent and manage diseases, injuries, and other conditions that threaten human helath. Keeping track of the number of people who acquire or have a particular health-related condition guides the deployment of interventions, distribution of grant funding for research on particular diseases, and development of public health policy.

In the United States, the Centers for Disease Control and Prevention (CDC) is the arm of the federal government responsible for promoting and protecting public health. On the infectious disease front, the CDC receives reports on the occurrence of certain infectious diseases, called **notifiable diseases**, from regions in the United States and its territories. The data received from state and local health agencies each week is compiled into a large searchable database called the National Notifiable Diseases Surveillance System (NNDSS) and published in the *Morbidity and Mortality Weekly Report* (MMWR), which is available in both print and electronic formats.

The data maintained within the NNDSS tables is available for retrospective analysis and also used to predict trends in disease occurrence in populations by time and place.

Two important measurements of disease occurrence and distribution are **morbidity** (illnesses due to a disease) and **mortality** (deaths due to a disease). The morbidity of a specific disease is defined as the number of susceptible people in a population that have the disease during a specific period of time, and is usually expressed as a rate. Mortality may also be expressed as a rate, and reflects the number of deaths due to a particular disease in a population over time.

Frequency of Disease in a Population

The frequency at which a disease occurs in a population is a way to assess risk and disease impact. One way to measure disease frequency is to simply count how many people are afflicted with it in a given period of time. However, using simple counts prevents comparison among populations, which may vary vastly in size. Therefore, disease frequency is usually expressed as a proportion of the number of people affected by the disease to the population size, over a specified time period.

Two specific statistical measures widely used in epidemiological investigations are **incidence** and **prevalence**. Incidence is a measure of the number of NEW cases of a disease during a specific time period. Incidence is used as a way to understand risk factors, such as the cause of a health-related event or concern for disease spread. Prevalence refers to the total number of both new and existing cases in a population over time, and provides an indication of the overall health of the population during a time period.

Both of these statistics are measures of disease over time. For this reason, they are often expressed as a rate:

Incidence rate = Number of new cases of a disease in a population ÷ Number of at-risk people during a time period

Prevalence rate =Number of cases of a disease in a population ÷ Number of at-risk people during a time period

Because the number of cases of any disease may be small, and the size of the population under study may be very large, the resulting number may be so exceptionally small that it is perceived to be of no consequence. Therefore, these measures are often expressed as a percent, or multiplied by a factor of 100, 1,000, or even 100,000 so that the rates are expressed in number of people per 100, 1,000, or 100,000 individuals, respectively. For example, if over the course of one year, five women in a study population of 200 women (5/200) develop breast cancer, then the calculated incidence of breast cancer in this population is 0.025. Such a small number might lead some people to presume their disease risk is also small. Therefore, the incidence may be expressed as a percent (2.5%), or a multiplier can be used to express the disease rate as 25 breast cancer cases per 1,000 women per year.

Prevalence estimates the likelihood that someone in a group will have a disease, and is often used as an indicator of the overall healthcare burden of a disease. Prevalence is highly dependent on the duration of the morbidity associated with the disease. The prevalence of chronic diseases will continually increase as the cases accumulate over time since it is a measure of both new and existing cases.

For example, a survey asking about personal experience with colon cancer was provided to 80,000 people, with 2,400 responding that they had been recently diagnosed with the disease, and 7,000 people responding that they'd had the disease for more than a year. The prevalence rate for colon cancer in this population can be determined by adding new and existing cases (9,400) and dividing by the size of the population (80,000). Therefore, the prevalence of cancer in this population is 0.1175, which can be expressed as 11.75%, or 118 colon cancer cases per 1,000 people.

Incidence and prevalence are two fundamentally different statistics. Keeping track of new cases of a disease requires an extensive network of reporting, while prevalence can be determined by surveying members of a population at a given point in time. Although there are limitations, if the disease is fairly stable in the population, has an average time of duration, and is not irreversible, incidence can be estimated using the prevalence data, and vice versa, using the following relationships (where Time refers to the average amount of time a person is sick with the disease):

Prevalence rate = Incidence *x* duration (in days, weeks, or months) of the disease

Incidence rate = Prevalence / duration (in days, weeks, or months) of the disease

Example: A prevalence survey conducted in upstate New York in 2013 revealed that 200 people in a study population of 16,000 Saratoga County residents were diagnosed with anaplasmosis, a bacterial disease transmitted to humans by ticks. For appropriately treated patients, the average amount of time that a person is sick with this disease is approximately four weeks.

1. What is the prevalence of anaplasmosis in Saratoga County, expressed per 1,000 people?

2. What is the estimated **annual** incidence of anaplasmosis per 1,000 Saratoga County residents? (Hint: Because this asks for the annual incidence, time should be expressed in years.)

3. In 2013, there were 223,865 people living in Saratoga County. Therefore, how many of those people would be expected to have anaplasmosis in 2013?

Measures of Association

Measuring the frequency of health-related events in populations is a useful way to assess and compare the health status of people in a population at one time, at different times, among subgroups of the population, or between populations. However, knowing how frequently a disease occurs in a single group does not indicate whether being a member of that group increases a person's risk of experiencing a specific health-related event.

Therefore, identifying the cause of a health-related event in epidemiology usually includes comparing disease rates between groups of people who differ by exposure. By measuring and comparing the frequency of health related events between groups where one is exposed and one is not, it is possible to evaluate if there is an association between a particular risk factor (such as smoking) and a positive or negative impact on health (such as cardiovascular disease).

For **cohort** studies which involve a group of people who share the same experiences, epidemiologists may make comparisons of disease frequency by calculating ratios of the variables. The **risk ratio** (also known as **relative risk**) gives an indication of the **strength of the association between a factor and a disease or other health outcome**. To calculate the relative risk, the incidence of the health-related event in a group that was exposed to the condition or variable is divided by the incidence of the same variable in the group that was not exposed. In general, a calculated risk ratio equal to or close to one indicates that there is no difference in risk, because the incidence is approximately equal in both groups. Ratios greater than or less than one suggest higher or lower risk, respectively.

To calculate relative risk in a study involving a cohort, the conventional method is to organize the data in a format known in statistics as a "2 x 2" table. An example is shown in Table 1:

Table 1. Standard 2 x 2 table for relative risk calculation.

	Outcome			
	Yes	**No**	**Total**	**Incidence of outcome**
Exposed	16	108	124	16/124 = 0.13
Not Exposed	14	341	355	14/355 = 0.04

Relative risk is calculated by dividing the incidence of the health event for the exposed group by the incidence of the health event in the unexposed group:

RR = incidence of outcome in exposed group / incidence of outcome of non-exposed group

RR = 0.13/0.04 = 3.25

In this case, because the calculated value is more than one, there is an increased risk associated with exposure to the risk factor. Specifically, the people in the exposed group were 3.25 times more likely to have the health event than those in the non-exposed group.

Example: To determine if patients who take prophylactic antibiotics before surgery are more or less likely to develop a hospital-acquired infection (HAI) of the wound, two groups of surgery patients were compared. One group with eighty participants took an antibiotic prior to surgery, and a second group of seventy patients did not take the antibiotic. Six people in the antibiotic group developed an HAI after surgery, and nine people in the no antibiotic group ended up with an HAI. Calculate the relative risk for this health-related event.

Table 2. Relative Risk Example

	Outcome			
	HAI	**No HAI**	**Total**	**Incidence of outcome**
Antibiotic	6	74	80	6/80 = 0.075
No antibiotic	9	61	70	9/70 = 0.13

RR = incidence of HAI for exposed group / incidence for non-exposed group

RR = 0.075/0.13 = 0.58

Because the relative risk is less than one, there is a reduced risk for a patient of getting a hospital-acquired infection if they are given an antibiotic before surgery. Specifically, someone who gets a pre-surgery antibiotic has 0.58 times the risk of an HAI, meaning that taking a pre-surgery antibiotic cuts the risk of HAI by almost half.

Another option to compare frequencies of health events is to calculate the **risk difference**, in which the difference between the two measures is determined by subtraction. The risk difference provides a measure of the **public health impact** of the risk factor and indicates how the health event might be prevented if the risk factor were eliminated.

The cohort study above examined if prophylactic antibiotics reduced the risk of getting a hospital acquired infection for patients. Note that the incidence of HAI in the antibiotic group was 75 per 1,000 people, and the incidence of HAI in the no antibiotic group was 130 per 1,000. The difference between these two values (55) indicates the number of HAI cases that could be prevented through prophylactic antibiotics before surgery. In this case, HAI would be prevented for 55 people (per thousand) if they are given an antibiotic before surgery.

Example: To determine if people who take a proton-pump inhibitor to combat heartburn are more or less likely to develop gastroesophogeal reflux disease (GERD), two groups of patients were compared. One group with 43 participants took the PPI daily, and a second group with 39 patients did not. After 3 months, 6 people in the PPI group developed developed GERD, while 5 people in the no PPI group developed GERD. Calculate the risk difference and indicate whether taking a PPI reduces the risk for GERD.

Using a case-control (as opposed to a cohort) study, relative risk is also a way for epidemiologists to track risk factors associated with disease outbreaks and potentially assign a cause, such as during a sporadic outbreak of a food-borne disease.

Example: On February 12, 2014 a forty-three-year-old man in New York was hospitalized with a one-week history of diarrhea and vomiting followed by fever, neck pain, and headache. This was the first reported (index) case of a sporadic outbreak of listeriosis, a disease caused by the bacterium *Listeria monocytogenes*. Almost everyone who is diagnosed with listeriosis has an invasive infection, meaning that the bacteria spread from their intestines to their bloodstream or other body sites, including the central nervous system.

An epidemiological investigation of this event identified 630 laboratory confirmed listeriosis cases across 11 states. To identify the source of the bacteria, a **case-control** study was conducted to compare the foods eaten by 52 of the patients with confirmed **cases**, with a group of 48 healthy **controls** who were matched to the case patients by gender, age, and geographic location. All 100 people were asked to complete a questionnaire about the foods they had eaten just prior to the index case report. The data is illustrated in Table 3.

Table 3. Questionnaire data				
	Ate food		**Did not eat food**	
Food item:	Sick	Not Sick	Sick	Not Sick
Weiner brand hot dog	24	28	22	26
Raggle brand sausage	20	32	29	19
Dairydelish yogurt	38	14	13	35
Yummyum ice cream bar	28	24	23	25

So… which food was contaminated? Calculate the relative risk for each food, and the highest number wins. Start by calculating the incidence for each group (first food item is shown):

Weiner hot dogs	Incidence exposed	24/52 = 0.46	RR:	0.46/0.5 = 0.92
	Incidence not exposed	22/48 = 0.5		
Raggle sausage	Incidence exposed		RR:	
	Incidence not exposed			
Dairydelish	Incidence exposed		RR:	
	Incidence not exposed			
Yummyum	Incidence exposed		RR:	
	Incidence not exposed			

Based on your calculations, which food is associated with this food-borne outbreak of *Listeria*?

Epidemiology Problem

On July 30, 2013, the New York State Department of Health received a complaint from a person who said that he and his entire family had become very ill with vomiting and diarrhea after eating at a particular restaurant. He went on to say that his two-year-old son, Devin, became so dehydrated that he required hospitalization. After rehydration therapy, Devin was well enough to return home. A specimen taken from Devin's stool was cultured on several types of media, including Sorbitol-MacConkey (SMAC) Agar, Salmonella-Shigella (SS) Agar, and Mannitol Salt Agar (MSA). Pink colonies grew on the SMAC plates, but no colonies appeared on the MSA plate. Pertinent results and additional tests are provided in Table 4, or your instructor may provide you with the actual media containing the cultures you should use in this analysis.

Table 4. Laboratory results for bacteria cultured from stool specimen.				
Gram stain	**MacConkey Agar**	**Catalase**	**Oxidase**	**TSI**
Pink single bacilli	Colonies were translucent and beige colored	Bubbles formed when H_2O_2 was added	No color change when smeared on a DrySlide Oxidase card	K/A H_2S gas

Based on the laboratory results, what bacterial genus is the most likely cause of Devin's illness?

____

What is the name of the gastrointestinal **disease** caused by infection with this bacterium?

____

Over the next 10 days, the hospital where Devin had been treated saw an additional 19 cases of rapid-onset gastroenteritis in people who dined at the same restaurant as Devin and his family. The Department of Health initiated an investigation, which included interviewing restaurant staff and people who had eaten there at some point over the previous two weeks. Samples of food taken from the restaurant at the time of the interviews did not test positive for any harmful bacterial agents.

To determine what food item might have been contaminated, a case-control study was conducted with the 19 people who developed food poisoning after dining at the restaurant matched with 20 controls, who had eaten at the restaurant but did not get sick. The responses were compiled and the data is shown in Table 5.

Table 5. Data from the case-control study				
	Exposed		**Not Exposed**	
Food item:	Sick	Not Sick	Sick	Not Sick
Hamburger	8	11	9	11
Hot dog	7	12	8	12
Fried chicken	9	10	12	8
French fries	10	9	11	9
Potato salad	16	3	4	16
Soda	11	8	11	9
Water	9	10	6	14
Beer	10	9	10	10

Can a particular food item be associated with the occurrence of disease among the people that ate at the restaurant? If yes, which food?

____

Epidemiologists were interested in knowing if this was a sporadic outbreak or an indication of a disease becoming more common in the upstate New York region. Therefore, a quick analysis was performed by comparing the incidence of this disease at 4 times throughout the year of 2013, at weeks 12, 26, 40, and 52. Retrieve this data from the NNDSS database (you can find it at cdc.gov → MMWR → State Health Statistics → NNDSS Tables → Search Morbidity Tables).

Week 12: Number of reported cases _________

Week 26: Number of reported cases _________

Week 40: Number of reported cases _________

Week 52: Number of reported cases _________

In this case, the size of the population would be considered the same for each of the weeks, therefore it is possible to compare the number of reported cases without calculating the incidence. From this data, what can you conclude overall about the occurrence of this disease in upstate New York? Is there any indication that we are on the verge of an epidemic of this disease?

____

____

____

Blood: The Good, the Bad, and the Ugly

Hematopoiesis

Blood is a fluid that transports and delivers nutrients and oxygen to body cells, and removes metabolic waste to be excreted. In vertebrate animals such as ourselves, blood is composed of several different types of cells (the cellular components) suspended in a watery liquid called plasma which contains dissolved solutes and proteins (the humoral components), including immunoglobulin proteins called antibodies. Both the cellular and humoral components of blood are transported throughout the body via the cardiovascular and lymphatic systems.

Cell-mediated immunity

The production of blood cells is called **hematopoiesis** and occurs throughout our lifetime, primarily in the bone marrow and in the lymph nodes after birth. Blood cells originate as "stem cells," which become committed to differentiate into mature cells along one of two different paths: **myeloid** or **lymphoid**.

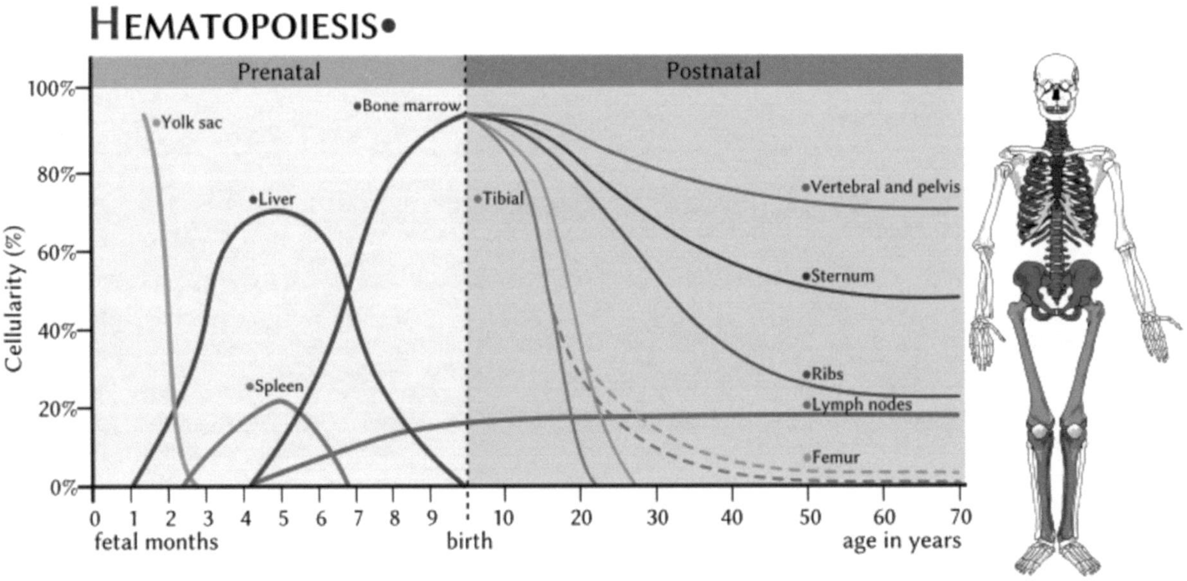

Figure 1. The production of blood cells in hematopoiesis.

Found most abundantly in blood are the **erythrocytes** (red blood cells), which derive from the myeloid cell line. These are highly specialized cells filled with hemoglobin designed to transport respiratory gases to and from the lungs. Another type of cell that derives from the myeloid line are megakaryocytes, which produce **thrombocytes** (platelets).

Leukocytes (white blood cells) are our primary cellular system of defense against disease. White blood cells circulate and patrol the blood and tissues, where they encounter and examine foreign entities invading the human body, sometimes engaging in mortal combat.

Leukocytes are derived from either the myeloid or lymphoid cell lines. In general, the cells in the myeloid cell line play a key role in the **innate** (also called "nonspecific") immune response. Myeloid white blood cells include **granulocytes** (neutrophils, eosinophils, and basophils) and **monocytes.** These are primarily phagocytes that hunt down, engulf and destroy invading entities, while sending messages to other types of immune cells that an invasion has taken place. The innate response also includes humoral components, such as complement. The majority of bacterial infections that breach the security of our skin and mucous membrane barriers are dealt with swiftly and effectively by our innate immune response.

Cells in the lymphoid cell line, called **lymphocytes**, are activated via signals received from myeloid cells. Lymphocytes launch the **adaptive** (also called "specific") immune response. Lymphoid cells differentiate into natural killer (NK) cells, T-cells and B-cells. When activated, B-cells differentiate into plasma cells which produce antibodies.

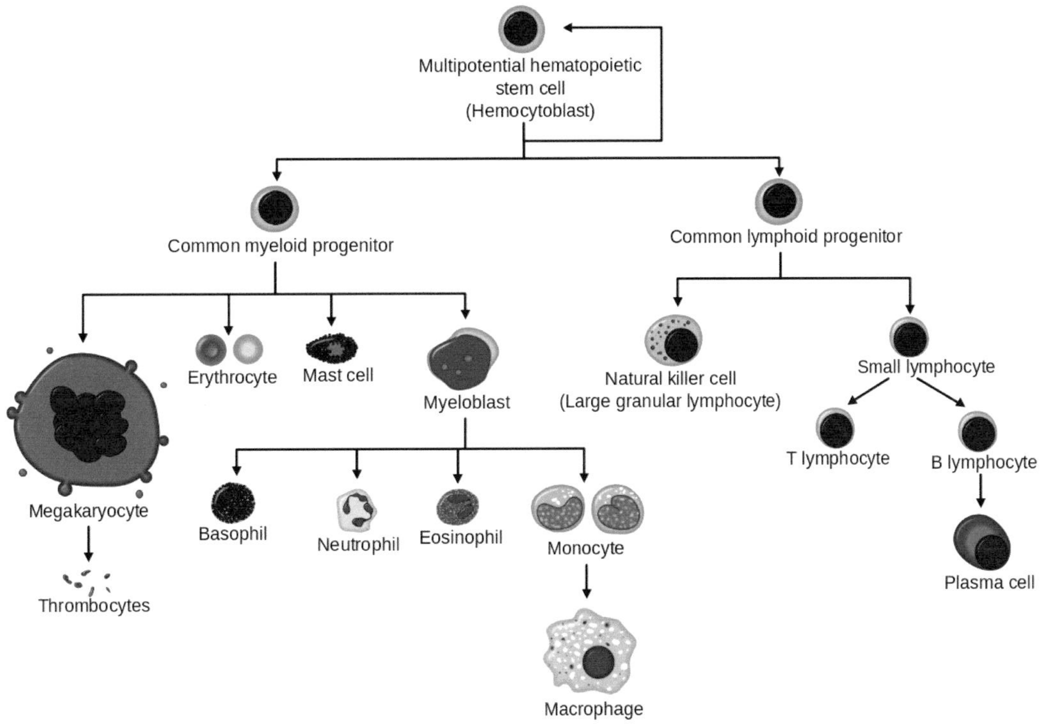

Figure 2. Differentiation of hematopoietic stem cells into various blood cell types.

Cells involved in the adaptive immune response	Cells involved in innate immune response

Humoral immunity

The term "humoral" refers to the liquid portion of the blood. Although both innate and adaptive immune responses have humoral components, use of this term is most closely associated with the production of antibodies as the culmination of the adaptive immune response.

Antibodies (also called immunoglobulins) are proteins that recognize and bind specifically to foreign structures associated with cells or objects, which are called **antigens.**

Antibodies have unique molecular structures designed to match with only one type of antigen. Antibodies bind to and neutralize antigens in several ways.

Antibodies are produced by B-cells after they are activated by signals from other immune cells. Activated B-cells undergo rapid growth and produce a clone of cells all producing one specific type of antibody molecule.

A subpopulation of the activated B-cells develop into memory cells, which "remember" the antigen. Memory cells are primed and ready if the antigen is encountered again. This provides the biological basis for vaccination, in which the immune system is artificially exposed to an antigen without triggering disease. Antibodies and memory cells are produced as if the exposure was to the actual antigen, and this provides future protection against the actual disease.

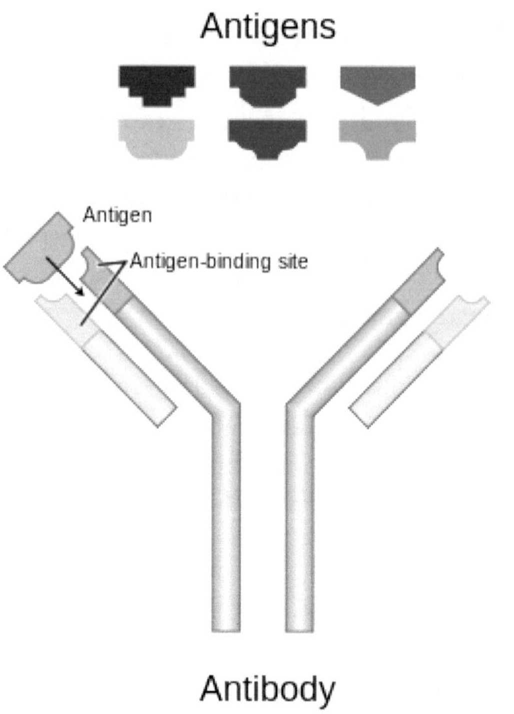

Figure 3. The structure of an antibody molecule.

Antibody molecules persist in the blood and may continue to circulate for months to years after the antigen exposure occurred. Both the antibodies, and microbial antigens themselves, can be detected and measured in laboratory tests called immunoassays, which can aid in the diagnosis of an infectious disease.

The overall human immune response is exquisite and enormously complex, and is actually a course unto itself. Characteristics of each of the two "arms" of the human immune system are summarized in Table 1. Our goal in microbiology is to better understand the types of interactions that occur when the immune system encounters harm-causing microbes, but it's also important to understand immunity from the perspective of the home team (our commensals) as well.

Table 1. The two arms of the human immune system.

Components of the Human Immune System	
Innate Immune Response	**Adaptive Immune Response**
Response is non-specific	Specific for pathogen or other antigen
Immediate maximum response	Lag time before maximum response
Cellular and humoral components	Cellular and humoral components
No immunological memory	Exposure leads to immunological memory
Found in nearly all forms of life	Found only in jawed vertebrates

Antibodies as an indirect indicator of an infectious disease

Serological tests for antibodies or antigens in blood are widely applied in clinical laboratories because they provide evidence of infection. Detecting antibodies in a patient sample is not necessarily a direct indication that the person has a disease, but rather shows past or present exposure to the disease-causing agent.

Infectious mononucleosis (IM), a disease that may follow infection with Epstein-Barr virus (Human herpesvirus 4) or Cytomegalovirus (Human herpesvirus 5), is an example of a fairly common disease often diagnosed on the basis of detection of antibodies in a person's blood. One type of immunoassay for IM detects "heterophile" antibodies in a patient's serum. These are weak, early, broad specificity IgM class antibodies produced against poorly defined antigens, which happens to include cow, pig, and horse red blood cells. The reaction between heterophile antibodies and animal red blood cells results in "hemagglutination," or clumping of the red blood cells, which can be visibly observed.

For whatever reason, people with infectious mononucleosis and a few other infectious diseases (hepatitis and rubella) have elevated levels of heterophile antibodies. They sometimes cross-react with "self" antigens, and may also be found in people with autoimmune disease.

Use reliable internet sources to research the clinical signs and symptoms consistent with a diagnosis of infectious mononucleosis (IM) and compile a list below.

____

____

____

____

We will be performing a commercially-available qualitative assay for detection of heterophile antibodies as an indicator of infectious mononucleosis. Your instructor will demonstrate how to perform the test, using positive and negative controls.

The method below corresponds to the Fisher HealthCare Sure-Vue Color Mono Test kit, but other kits may be substituted by your instructor. The Sure-Vue test kit uses specially treated horse red blood cells. If heterophile antibodies are present in the patient sample, the red blood cells will agglutinate which will appear as dark clumps against a colored background.

Clinical Scenario

A 16 year-old female patient reports symptoms that include a sore throat and feeling achy and overly tired. Clinical observations include a fever of 102°F and swollen lymph nodes in the neck region. The initial differential diagnosis made by the physician includes streptococcal infection (strep throat) and infectious mononucleosis. A blood sample is taken and tested using the Sure-Vue Mono Test.

To perform the test, you will need to obtain a test card, one patient serum sample, and a bottle of test "reagent" (a suspension of horse erythrocytes) from your instructor.

Place one drop of the patient sample inside the circle on the card. Shake the reagent bottle, and add one drop of the reagent next to the drop of patient sample.

Using a wooden toothpick, mix the two drops together thoroughly so that the combined drops completely cover the surface within the circle.

Rock the slide back and forth gently for 1 minute, then set it down on the lab bench and let it sit undisturbed for an additional 1 minute. Without moving the slide again, look at the circle to see if clumping is visible.

> A positive reaction will have moderate-to-large sized dark clumps against a blue-green background, distributed uniformly over the surface of the test circle.
>
> A negative result will show no clumping, although there may be a slightly graining appearance, against a greenish-brown background.

Was your patient sample positive or negative for this test? ________________

Based on the clinical signs and the result of the test, can the doctor be absolutely certain that the patient is infected by either EBV or CMV, and that the infection is causing infectious mononucleosis? Explain your answer.

____

____

____

____

Blood cell counts as indicators of health and disease

One way to evaluate a person's (or animal's) state of health is to directly or indirectly examine cells or metabolites found in blood. In clinical laboratories, levels of various metabolites in plasma may be measured by chemical assay, and blood cells can be distinguished and counted using automated methods.

It is also possible to evaluate the cellular components in blood by directly observing them with a microscope. To be able to see individual cells, it is first necessary to create a very thin film of blood with a "feathered edge" (a single layer of cells) on a glass slide. Unstained and stained blood smears with a near perfect "feathered edge" are shown in Figure 4.

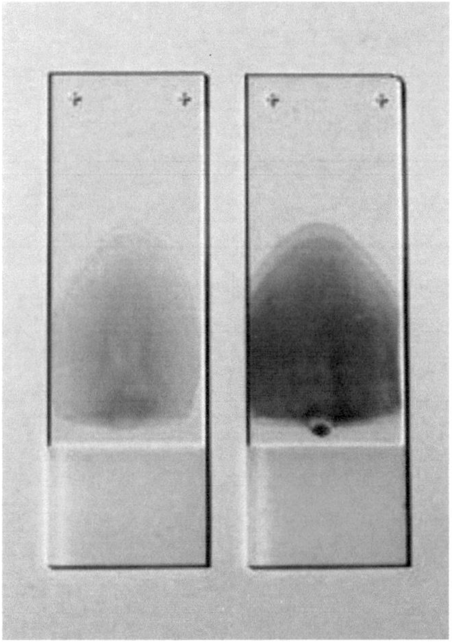

Figure 4. Stained and unstained blood smears.

Once the blood smear is stained, the cells are visually inspected with a microscope. One of the most commonly used differential stains is the Wright-Giemsa stain, which stains red blood cells a pinkish-red color, and stains the nucleus and cytoplasm of white blood cells various shades of purple. Stained blood smears are examined to evaluate the appearance of the blood cells, and to count the number of different types of white blood cells present. Blood smears may also be examined to see if the blood contains any protozoal or bacterial pathogens associated with disease.

From an infectious disease perspective, the number of white blood cells and the relative percentages of different types of cells may indicate whether a person has a disease. This can be noted as a departure (either higher or lower) from established "normal" values.

Normal ranges, expressed as a percent of total white blood cells, are provided in Table 2—note that these values vary across age and gender and are therefore only approximations provided for the purpose of this lab. The absolute number of white blood cells generally considered "healthy" ranges from 3.5 to 11 x 10^9 cells/L.

Table 2. Normal ranges for white blood cells in peripheral blood.

White blood cell type	Range of relative values for "normal"
Neutrophils	50–75%
Lymphocytes	15–35 %
Monocytes	3–10%
Eosinophils	1–7%
Basophils	0–2%

Deviations from these "normal" values can be an indication that an active infectious disease or a blood-associated disorder is ongoing.

For the types of white blood cells listed in the table below, research what disease conditions are associated with a relative increase or decrease in the numbers of that particular cell type:

White blood cell type	Conditions associated with a relative increase	Conditions associated with a relative decrease
Immature neutrophils, in which the nucleus looks like a single "band"		
Lymphocytes		
Eosinophils		
Basophils		

To determine the relative percentages of the different types of white blood cells found in a person's blood, it's important to first know what each cell type looks like and be able to tell them apart. These will be shown in the lab and/or provided as handouts before you start your investigation.

The Differential White Blood Cell Count (a "Diff")

When a differential cell count is performed in a clinical lab, the technologist first makes a blood smear either manually or by machine, hoping to achieve that perfect feathered edge. Smears are routinely stained with Wright-Giemsa stain, and this is the staining method used to prepare the slides we will be viewing in this lab.

After staining, the smear is observed with a light microscope using the oil immersion objective lens. An initial scan to evaluate the appearance of the red blood cells and platelets is performed, followed by a more detailed analysis of the populations of white blood cells present.

To determine the relative number (percent) of each type of white blood cell, the smear is scanned using a pattern that prevents the observer from counting the same cells more than once. As each white blood cell is encountered, it is identified by cell type and recorded. A "tally" is kept and when the total number of cells observed reaches 100 the proportion of each type can be easily determined as the number of that cell type/100. The percent is calculated by multiplying by 100.

Obtain a prepared slide of a Wright stained blood smear from the front bench. Using the scanning pattern illustrated in Figure 5 and as demonstrated, scan the entire slide. Identify and count each white blood cell you see, until you have reached 100 cells.

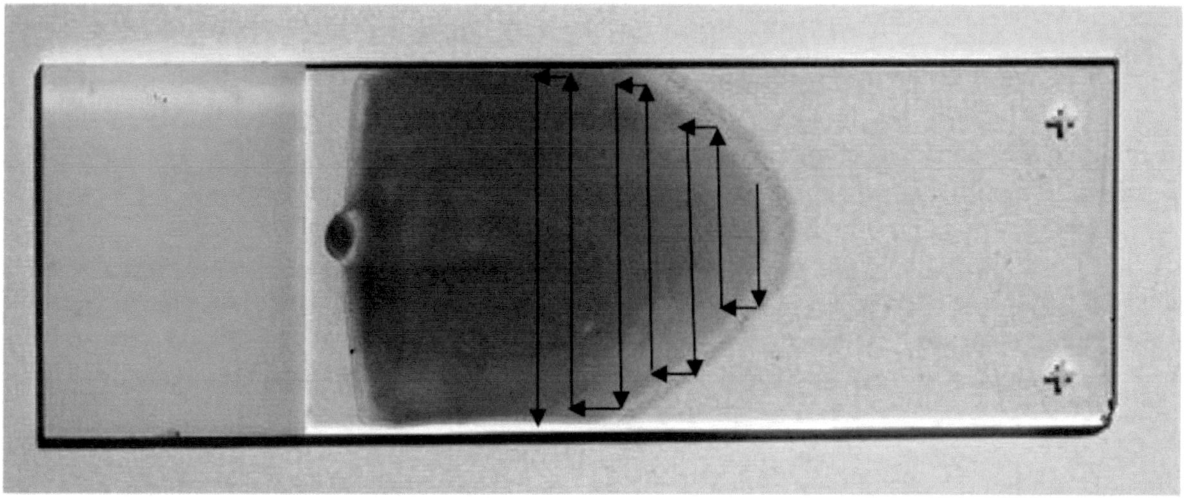

Figure 5. Scanning pattern for microscopic evaluation of blood smears.

Using a cell counter (if one is available in your lab) makes the tally process a little easier. Record your results in the table below:

White blood cell type	Number of cells	Percent
Neutrophils (including bands)		
Lymphocytes		
Monocytes		
Eosinophils		
Basophils		

Based on the results of your differential cell count, provide an analysis in terms of how the blood compares to the "normal" values. What do the results suggest about the overall state of health of the patient? Provide at least three specific examples from among your results to support your opinion.

____

____

____

____

Staining methods to detect pathogens in blood

Although blood is generally thought of as a "sterile" tissue, some viruses, bacteria, protozoa and helminthes may actually live in blood cells and circulate throughout the human cardiovascular system. Infections of the blood can range in severity from a transient bacteremia (bacteria in the blood) to septicemia, a "septic" state which may result in severe illness and death.

Invading microbes in blood are usually attacked, destroyed and removed by the immune system as quickly as possible. However, we know from research that some highly adapted microorganisms may survive and even reproduce within blood cells. To accomplish this feat, the microbes must "convince" the immune system to leave them alone by either evading or subverting the innate or adaptive responses.

Although parasitic blood infections such as malaria and heartworm are relatively well known, it is becoming increasingly obvious that bacteria also cause long term infections that can lead to chronic disease. Two examples of bacteria that infect and reproduce inside blood cells include *Rickettsia* spp. (Rocky Mountain Spotted Fever) and *Bartonella* spp., (best known for causing a disease referred to as "Cat Scratch Fever"). Both of these bacteria and several others are zoonotic diseases that have "spilled over" into humans. Most are transmitted to humans by insect vectors such as fleas, lice, and ticks.

To survive in blood, bacterial pathogens must first evade capture and destruction by phagocytic cells. One successful approach is for the pathogen to infect and then reproduce inside the very cells that were sent to destroy them. This and other strategies used by bacteria are referred to as "stealth pathogenesis." Attack strategies employed by stealth pathogens are fundamentally different from those used by the better known "frontal" pathogens, whose major tactic is to breach the host perimeter, reproduce and spread as quickly as possible before the immune system can rally for a response. Table 3 compares and contrasts these two strategies for host invasion and infection.

Table 3. Characteristics of Frontal versus Stealth Pathogens

	Frontal Pathogens	Stealth Pathogens
Incubation period	Short (hours to days)	Long (months to years)
Symptoms	Acute	Chronic
Immunity	Sterilizing	Non-sterilizing
Transmission	Direct	Indirect (vector)
Reproduction	Rapid	Slow
Carrier state	Uncommon	Common

Note. Adapted from "Front and stealth attack strategies in microbial pathogenesis" by D. S. Merrell and S. Falkow, 2004, *Nature, 430,* p. 250-256. Copyright 2004 by Nature Publishing Group.

Using stealth tactics, some microorganisms are able to invade their hosts, disseminate to parts of the body that have poor immune surveillance, evade phagocytosis and disarm the adaptive immune response. Many stealth pathogens that infect blood cells also have the ability to infect multiple types of animals, and are transmitted among hosts by blood-sucking insects.

We will be looking for evidence of infection by stealth microbes that reproduce inside of various blood cell types, by examining Wright-Giemsa stained blood smears.

Anaplasma spp.

Anaplasma spp. are stealth bacteria that invade red and white blood cells. Examples include *A. centrale* in cows, which invade red blood cells, and *A. phagocytophilum* in humans and other animals, which reproduce inside phagocytic white blood cells.

Although bacteria are presumed to have relatively simple lives, *Anaplasma* and other stealth bacteria have complex life cycles. Humans and animals are infected through the bites of a tick, which carry *Anaplasma* as well as a mélange of other disease causing agents, including *Borrelia* (Lyme disease), *Ehrlichia* (ehrlichiosis), and viruses that cause encephalitis in some patients.

Once in the blood, *A. phagocytophilum* binds to surface proteins on neutrophils and other granulocytes. After phagocytosis, the bacteria send signals that inhibit the development of a phagolysosome, and then are free to reproduce inside the endocytic vacuole (called the endosome).

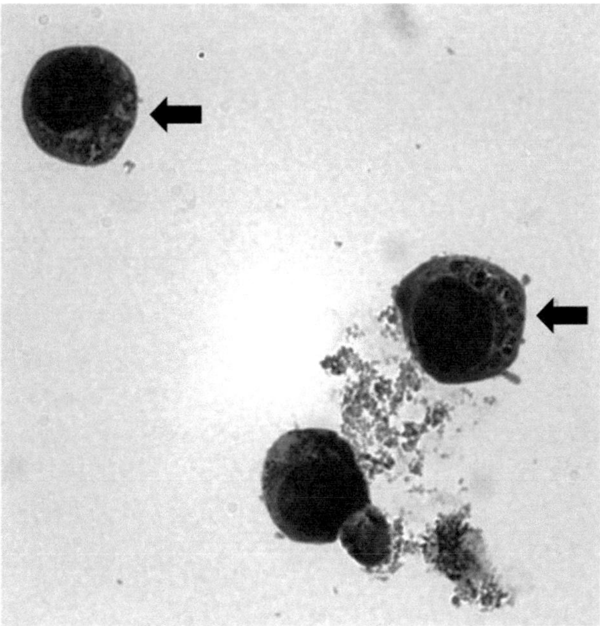

Figure 6. Anaplasma morulae in the cytoplasm of white blood cells.

This forms an observable structure, called the **morulae,** inside the infected cell (indicated by arrows in Figure 6), which can be seen by a careful observer on a Wright-Giemsa stained blood smear.

Obtain a stained blood smear slide labelled *Anaplasma*. Scan the slide using the oil immersion objective lens and when you encounter a phagocytic cell type, look closely at the cytoplasm and determine if there are morulae, which will stain light purple against the darker blue of the cytoplasm.

Keep a tally of the number of phagocytes you observe, both with and without morulae. Record your results in the table below:

Morulae of *Anaplasma* observed?	Number of cells with morulae	Number of cells without morulae	Percent of infected phagocytes

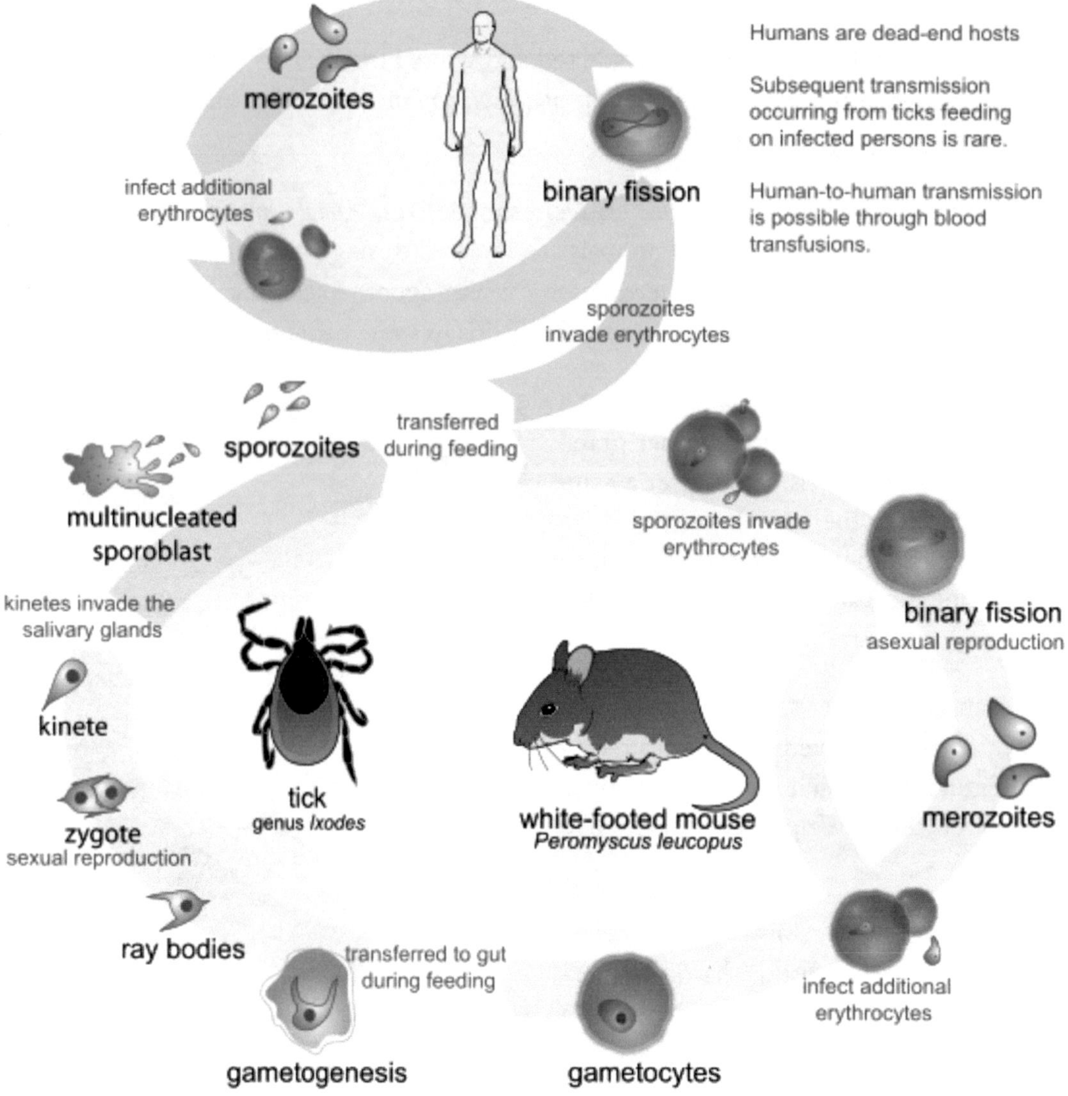

Figure 7. The complex life cycle of Babesia.

***Babesia* spp.**

Babesia spp. are protozoa classified as Apicomplexa, a group that also includes the better known *Plasmodium*, which causes malaria, and *Toxoplasma*, which causes toxoplasmosis.

Babesia has a complex life cycle that includes more than one life stage cycling through multiple hosts, as illustrated in Figure 7. Ticks introduce **sporozoites** into their host when they attach to take a blood meal. Sporozoites invade red blood cells and develop first into **trophozoites**, then **merozoites**, which are released and infect nearby red blood cells to perpetuate the cycle.

The trophozoites and merozoites can be observed both inside and outside of the cells in a Wright-Giemsa stained blood smear. Inside the red blood cells, the developing trophozoites take the shape of "rings" or "crosses" that stain purple and generally stand out distinctly from the red cell background, which is a pinkish color. The merozoites appear as amorphous blobs with a dark purple dot, which is the nuclear material of the parasite.

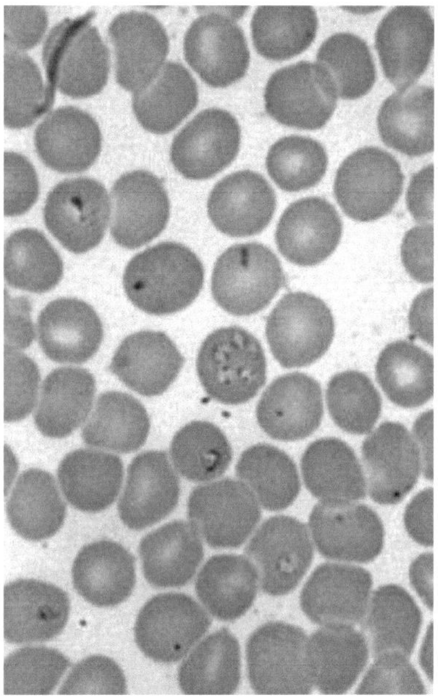

Figure 8. Babesia trophozoites in red blood cells.

Obtain a stained blood smear slide labelled *Babesia*. Scan the slide using the oil immersion objective lens, and look for evidence of *Babesia* infection, both inside and outside the red blood cells. Move to the thinnest part of the smear, where you can see a single layer of the red blood cells. Start counting the number of red blood cells, keeping track of the cells with obvious signs of trophozoite development and any extracellular merozoites. Continue to count until you've counted 100 cells. Use this data to estimate the level of parasitemia.

Number of red cells with *Babesia* trophozoites	Number of merozoites observed	Estimated level of parasitemia (trophozoites/merozoites observed per 100 cells counted)

According to a scholarly source (and cite the source), approximately what percentage of people in the United States are infected with *Babesia*, but may be asymptomatic or experiencing mild or chronic non-specific symptoms?

____

Reflect on the implications this might have on the supply of blood for blood transfusions available in the United States.

____

____

____

Dirofilaria

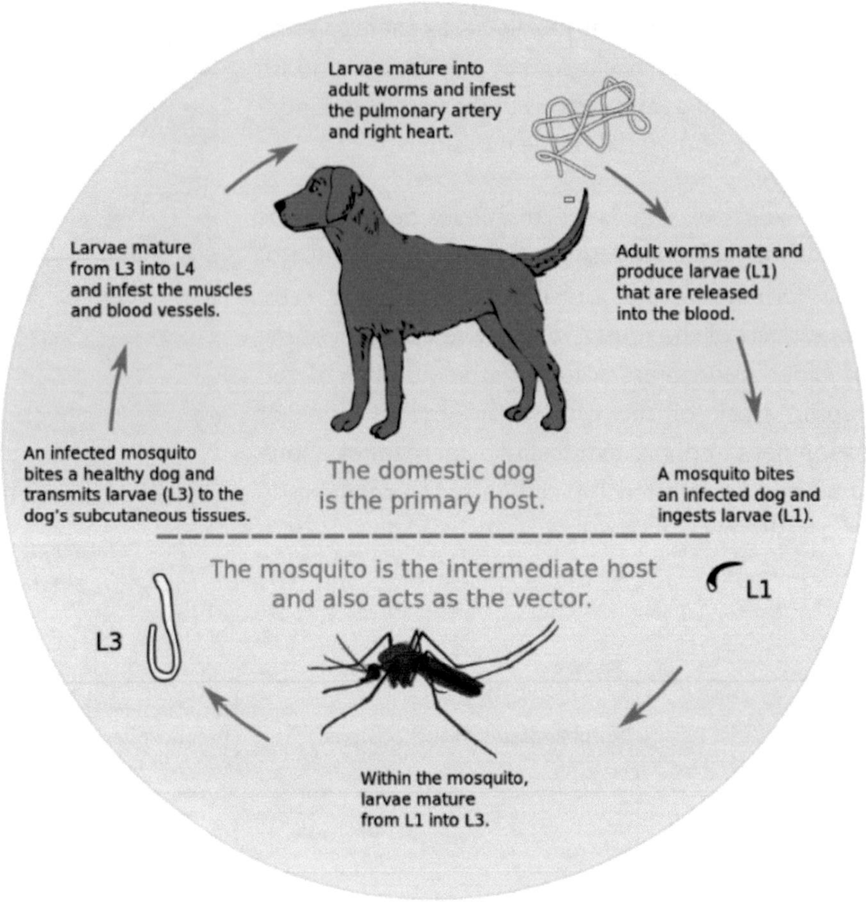

Figure 9. The complex life cycle of Dirofilaria.

Dirofilaria are nematodes (roundworms) also known as heartworms.

Dogs and cats are both known to be hosts for this type of parasitic worm, and heartworm control is a primary concern for pet owners and veterinarians alike.

Adult heartworms live in the right ventricle (one of the chambers) of the heart in dogs, and in the pulmonary arteries of cats. The adult worms produce **microfilaria** ("baby" heartworms) which can be seen in blood smears. Complications of an active heartworm infection in dogs range from nonexistent in early or mild infections to coughing, vomiting, trouble breathing and heart failure in advanced infections.

Obtain a stained blood smear slide of *Dirofilaria*. Scan the slide using the oil immersion objective lens, and look for the microfilaria worms which will be obviously larger than the blood cells. Figure 10 shows a picture of a single microfilaria in unstained blood as observed by bright field or phase contrast microscopy. In a Wright-Giemsa stained smear, the microfilaria look like worms, only stained blue and purple. Note the size of the microfilaria worm in relation to the red blood cells (which are approximately 10 μm in diameter). Adult heartworms may grow to be 20–30 cm long.

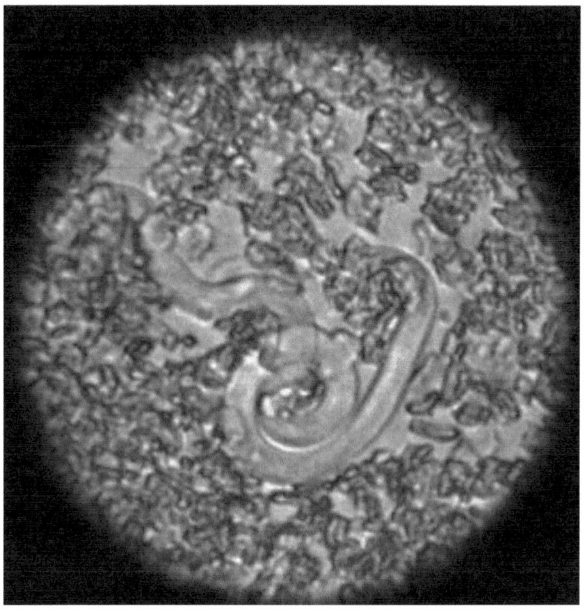

Figure 10. A microfilaria in a wet mount of whole blood.

Scan the entire slide, and keep track of the number of microfilariae you see.

How many individual microfilariae did you find in the blood smear? ____________________

Consider that the drop of blood from which the smear was originally made had a volume of less than 10 μl. Dogs are estimated to have approximately 80–90 ml of blood per kg body weight.

For a dog that weighs 100 lbs (45 kg), estimate how many microfilariae would be found in its circulation, using a value of 85 ml of blood per kg body weight.

____

____

____

Microbe Mythbuster

The great enemy of the truth is very often not the lie, deliberate, contrived and dishonest, but the myth, persistent, persuasive and unrealistic.

John F. Kennedy

What is Truth?

Truth is a philosophical construct whose meaning has been debated since humans invented language. That's not the focus of this endeavor.

This project is more about Reason, also a philosophical construct. Reason provides a path for pondering the truth. According to some, truth results when people apply reason appropriately about an issue at hand. This is the goal of science.

Maybe you have recently heard a claim about a nutritional supplement or seen an advertisement for a pharmaceutical drug touting amazing benefits if you take it, and wondered if you should. Or you thought about the health risks associated with getting a flu vaccine, or considered taking a probiotic because your cousin's friend said you should? How can you know what would be best for you?

There exists a vast body of scientific studies conducted on an infinite number of topics in science and medicine that is published in scholarly journals and stored in searchable databases. By conducting an organized review of the published research on the topic and applying "appropriate reason," you can decide for yourself what would be best for you, rather than relying on advice from ads or people you don't know.

The conduct of scientific research is guided by practices collectively referred to as the scientific method, in which experiments are designed to answer questions about a hypothesis. In a perfect world, experiments are carefully designed to ensure that the data collected and the results derived from them are objective and without bias. If the results are significant, the science gets published in a journal as a way to communicate the findings to other interested people. Volumes of journals have historically been stored in libraries, where articles contained therein could be read and copied if relevant. It is no longer necessary to hunt through dusty "stacks" of print journals to find a scientific article, because a huge number are now "open access" or available electronically through a library interface.

There are differences between articles published in scholarly journals and those in other types of publications, and the major difference is peer-review. It's important to note that use of the term "publication" includes papers published in electronic form as well as in print.

You should view a short video available at http://www.library.vanderbilt.edu/peabody/tutorial_files/scholarlyfree/, which explains how to tell the difference between a source reference from a scholarly publication and one published in the popular media.

Your instructor will be providing you with a microbiology-themed notion that you may have heard about before starting this project. Depending on the preferences of your instructor, you may investigate your assigned idea as a written assignment, or you may be asked to format the assignment in presentation software (such as PowerPoint) and make a formal presenation.

Before doing any research, reflect on and then write down your first impressions and personal views about the idea you've been provided. If you are unfamiliar with the idea, or even if you feel you understand it well, do a little background searching of the topic using popular sources (such as Google and Wikipedia) to gather background information before embarking on your scholarly search.

Debunk the Myths, Support the Truth

So much of what you hear on the evening news related to discovery in science and medicine comes from research conducted at universities and medical colleges. The funding for this research may come from government sources, and is therefore paid for by the taxpaying public. However, given the limited size of the pot, research is also conducted by private companies who then profit from research that culminates in a profit-bearing product. When research leads to publication in a "highly ranked" journal (ranked according to the journal's "impact factor," based on the number of times articles published in the journal are cited as a reference in other publications), a brief description of the study and its outcome are released to the popular media for reporting to the general public. Sometimes government policy is developed using published studies as a foundation for legislation.

Scholarly and non-scholarly reporting of scientific discovery means that people today have the unprecedented opportunity to make informed decisions about things that may affect their lives. However, it also provides fertile ground for the dissemination of information designed to "market" the idea to gain popular support. Once entrenched in the public conscience, misapplied "facts" may become "myths"—persistent, persuasive, and unrealistic. How do you tell the difference?

For this project, you will investigate whether a common microbiology idea is scientifically conceived and the degree to which it is "true," by evaluating and reporting on research published in scholarly journals. The components to be included in your report or presentation are specified below.

1. Review popular opinion and develop a thesis

Once you know your Mythbuster subject, look for background information and opinions among sources that are not considered "scholarly." This includes popular press sources such as newspapers, magazines, internet sources, or Great Aunt Martha who knows everything.

From your accumulated knowledge on the topic, develop a thesis on the topic, and assert what you think about it in a thesis statement—a one or two sentence prediction of what you believe to be true. The thesis statement should be focused and specific enough to be provable within the boundaries of your investigation.

As you search for the "reason" to back up the "truth," you may find that your thesis can't be supported by the available scientific evidence. However, you have to be flexible, objective, and honest when you construct and conduct your search of the scientific literature and not just look for ways to make your opinion seem true.

2. Search the scholarly literature

Scientists who think their research is significant communicate the results through publication in scientific journals. Most medical and scientific organizations publish journals related to a professional field—the American Society for Microbiology, for example, publishes several journals such as *Applied and Environmental Microbiology* and *Journal of Clinical Microbiology*, among others. Manuscripts submitted to scientific journals are sent to a panel of other scientists, who review them for scientific legitimacy and integrity. This insures that the data and results are obtained from carefully designed, reproducible experiments, and the conclusions are evidence-based. Once they are peer-reviewed and approved, they are incorporated into a volume of the journal and published.

It is important to consider that in a perfect world, using science and the scientific method to understand nature is a logical, objective, and totally unbiased process, that peer-reviewers are always honest, and that peer-reviewed articles represent the "truth." As several recent high profile cases illustrate, in which published studies have been "retracted" due to fraud on the part of the researchers and/or their reviewers, the process isn't perfect. This is particularly true when the financial or personal stakes are high.

Once you have developed your thesis statement, the next step is to look for published research studies pertaining to your topic. You can refer to http://www.wikihow.com/Find-Scholarly-Articles-Online/ for a concise overview of how to construct and conduct a search for scholarly articles on a topic of interest.

Many libraries at colleges and universities, such as the State University of New York library system, have access to huge databases containing millions of scholarly articles. Therefore, another excellent starting point is to enlist the assistance of a reference librarian in your college library, who can tell you what article databases are available and can help you construct your search. Reference librarians are particularly helpful when it comes to deciding on the right words or phrases, so that your search yields a manageable number of returns, not too few or too many.

Be objective when you decide on which articles to read further. Don't limit yourself to only those that agree with your thesis 100%. Peruse the abstract, and if it sounds like the article will be relevant to your idea, download the entire article (full text) and read the full content.

3. Create an annotated bibliography of selected scholarly articles

At this point you have (hopefully) browsed through a large list of articles pertaining to your subject. For those that you decided to read in greater depth, prepare a bibliography using the citation format preferred by your instructor. Some of the databases will actually write the citation for you, and again, your reference librarian can help you locate and access the citation application if it exists for that database.

You should provide citations for all of the articles you selected. Of those you include in the bibliography, select three of the articles that you feel exemplify your idea, and write a brief annotation to accompany the citation. "Annotated" means that after the citation, write a brief one to two paragraph summary of the objectives and outcomes of the research presented in the article. The final sentence of the summary should discuss how the article relates to your thesis. An example of an annotated reference is shown below (the citation format is APA).

Fava, F., Lovegrove, J. A., Gitau, R., Jackson, K. G., & Tuohy, K. M. (2006) The gut microbiota and lipid metabolism: Implications for human health and coronary heart disease. *Current Medicinal Chemistry, 13*, 3005-3021.

Summary: Coronary heart disease (CHD) is the leading cause of mortality in Western society, affecting about one third of the population before their seventieth year. This article reviews the modifiable risk factors associated with CHD and discusses the hypothesis that diets rich in sources of dietary fiber and plant polyphenols promote better coronary health. Plant fibers are metabolized by the gut microflora, and are converted into biologically active compounds that are complementary to human metabolism. Metabolism of plant fibers by the gut microflora may prevent or otherwise beneficially impact impaired lipid metabolism and vascular dysfunction that typifies CHD and type II diabetes. Overall this article supports my thesis that the bacteria in the human gut make positive contributions to a person's overall good health.

4. Write a summary and conclusion

Paper Option: In a paragraph (or two), summarize the scope of the project, the idea you are investigating, and restate your thesis. In two to four paragraphs, summarize the research that you discovered in your search of the scholarly literature, being sure to include the appropriate citation for each reference. In a final paragraph (or two), compare and contrast the non-scholarly information with what you learned from your search of the science, and discuss whether the scientific evidence was in support of your thesis, or if the evidence did not support your view. Consider whether you are sticking with your thesis or if you want to change it, and what amendments might be appropriate based on the scientific evidence.

Presentation Option: Using PowerPoint (or other presentation software), develop your report into a ten minute talk, which you may be scheduled to give as an oral presentation.

Appendix A

Equipment, Supplies, and Cultures

The following is a list of equipment, supplies, cultures, and media needed to support this laboratory experience. Specific bacteria may be replaced with other cultures according to individual lab preferences or requirements. Amounts are not provided, because numbers will vary by section size. Best outcomes will be achieved if students work alone for labs in which new skills are introduced.

General Lab Equipment

- Microscopes and immersion oil
- Bunsen burners or sterilizers
- Inoculating loops and needles (metal or sterile disposable plastic)
- Metal forceps
- Pipette pumps
- Incubators set for body temperature and room temperature incubation
- Biohazard disposal container(s)
- Autoclave
- Refrigerator

Maintained Stock of Laboratory Supplies

The following should be readily available to students during lab and open lab times:

- Glass slides
- Clear plastic flexible laboratory metric rulers
- Kimwipes and/or lens paper
- Sterile 10 ml glass pipettes
- Sterile swabs
- Sterile wood applicator sticks

Replaced and refilled as needed each week:

- Staining "kits," each kit containing stains in dropper bottles for Gram and Endospore stain-

ing procedures (Crystal violet, Gram's iodine, 95% ethanol, Safranin, and Malachite Green)
- Spray bottles with laboratory disinfectant
- 70% ethanol for disinfecting forceps and spreaders, with beakers
- 4–6 racks (160–240 tubes) of sterile glass culture tubes, empty
- 2 racks (80 tubes) of tubes with sterile distilled water for making smears
- 2 sleeves (80 plates) of TSA plates
- 1 rack (40 tubes) of TSA slants
- 2 bottles (100 ml) of TSB

The recommended nonselective growth medium for routine culture is Tryptic Soy (TS), but other types could easily be substituted.

Lab 1: Biosafety Practices and Procedures

- Principles of Biosafety Evaluation and Student Affirmation form to assess understanding and document biosafety training.

Lab 2: The Microscopic World

Prepared slides (purchased, stained and mounted):

- Rectal (or fecal) smear (Gram stain)
- Mouth smear (Gram stain)
- Yogurt smear (Gram stained)

Cultures:

- Yogurt, store bought, any brand
- Yogurt, freshly made (prepared within a week of lab)

To make yogurt: Combine a large spoonful of yogurt (or a starter culture such as YoGourmet) with steamed milk in a glass pint canning jar with a lid. Stir vigorously to mix, then incubate at 38–40°C overnight. Refrigerate when milk has thickened into yogurt.

Lab 3: Bacteriological Culture Methods

Cultures:

- Overnight plate cultures of *Micrococcus luteus* and *Enterococcus faecalis*
- Overnight broth culture containing two bacteria

For broth cultures: Grow broth cultures of bacteria separately, combine in one tube right before the lab begins.

- For BSL-1 only: *M. luteus*, *Micrococcus roseus* in equal proportions
- For BSL-2: any combination of bacteria. Try to use bacteria which grow at a similar rate, or proportions added to the mixed culture can be adjusted to account for growth rate (for

example, 2x as much Gram-positive to Gram-negative ratio)

Culture Media:

- Tryptic Soy Broth (or other nonselective general growth medium) in bottles
- Tryptic Soy Agar plates
- Tryptic Soy Agar slants in culture tubes
- Tryptic Soy deeps in culture tubes (semisolid agar: use 3.5 g agar per liter TSB)

Lab 4: The Environmental Isolate Project

Cultures:

- Bacteria cultured from skin and isolated in pure culture

Culture media (available in lab for the duration of the project):

- TSA plates and TSA slants for primary culture and pure culture maintenance

The media necessary to identify the genus and species of the isolated bacteria will vary each week. See media requirements and supplies for Growth Characteristics and Food Safety labs.

Lab 5: Differential Staining Techniques

Prepared slides (purchased already stained and mounted):

- *Flavobacterium capsulatum*, capsule stained (or other capsule-stained smear)
- *Corynebacterium diphtheriae*, methylene blue stain (or other stain for metachromatic granules)
- *Mycobacterium tuberculosis*, acid-fast stain

Cultures:

- Overnight TSA plate cultures of a Gram-positive bacterium and a Gram-negative bacterium (suggested are *Escherichia coli* and *Staphylococcus saprophyticus*)
- Overnight TSA plate cultures of two species of *Bacillus*, each with a different endospore appearance: suggested are *B. subtilis* OR *B. cereus* (with oval, central endospores) AND *B. sphaericus* OR B. *globisporus* (with spherical, terminal endospores)

Lab 7: Metabolism, Physiology, and Growth Characteristics of Cocci

Prepared slides (purchased, stained and mounted):

- *Neisseria gonorrhoeae*, Gram stain

Cultures:

- Overnight TSA plate cultures of *Staphylococcus aureus*, *S. epidermidis*, and *S. saprophyticus*, *Micrococcus luteus*, *Enterococcus faecalis*, *Streptococcus pyogenes*

Media and Test Reagents:

- Nitrate broth, in bottles
- Nitrate reagents A and B, stored at 4°C
- Urease broth, in bottles
- Triple Sugar Iron (TSI) agar slants
- Mannitol Salt Agar (MSA) plates
- Blood Agar Plates (BAP)
- Bile Esculin (BE) agar plates
- 3% hydrogen peroxide in dropper bottles (for catalase test)
- DrySlide Oxidase test cards (for oxidase test)
- Staphaurex (or other coagulase test kit) cards and reagent

Lab 8: Metabolism, Physiology, and Growth Characteristics of Bacilli

Cultures:

- Overnight TSA plate cultures of a *Bacillus* spp., *Corynebacterium xerosis*, *Pseudomonas aeruginosa*, *Escherichia* coli, a *Salmonella* spp., *Citrobacter freundii*, *Enterobacter aerogenes*

Media and Test Reagents:

- Nitrate broth, in bottles
- Nitrate reagents A and B, stored at 4°C
- MR-VP broth, in bottles
- Methyl Red reagent, and VP (Barritt's) reagents A and B, stored at 4°C
- Triple Sugar Iron (TSI) agar slants
- Sulfide-Indole-Motility (SIM) agar stabs
- Indole (Kovac's) reagent, stored at 4°C
- Simmon's Citrate Agar plates or slants
- 3% hydrogen peroxide in dropper bottles (for catalase test)
- DrySlide Oxidase test cards (for oxidase test)

Lab 9: Microbiological Food Safety

Cultures:

- A selection of food items, such as ground beef, bagged lettuce, eggs, chicken, etc. (which may be "spiked" with bacteria). Students may be encouraged to include items of interest that they transport to lab with them.

Media and Test Reagents:

- Plastic disposable culture tubes with caps (to facilitate cleanup and disposal)
- Tryptic Soy Broth (TSB) in bottles

- MacConkey Agar (MAC) plates
- Sorbitol-MacConkey Agar (SMAC) plates
- Mannitol Salt Agar (MSA) plates
- TSI agar slants
- Staphaurex (or other coagulase test kit) cards and reagent

Lab 10: Germ Warfare

Cultures:

- Overnight TSA plate cultures of *Staphylococcus aureus*, *S. saprophyticus*, *Escherichia coli*, and *Pseudomonas aeruginosa*

Media and other supplies:

- 0.85% saline, sterile, in bottles
- 0.5 McFarland turbidity standards
- Mueller Hinton Agar (MHA) plates
- Antibiotic disks with: novobiocin, cefixime, azithromycin, ciprofloxacin, tetracycline
- Sterile blank filter paper disks
- Samples of a variety of known and potential disinfectants and antiseptics, such as a commercial home product, hand sanitizer, antibacterial dish soap, mouthwash, products made with essential oils, honey, etc.

Lab 11: Epidemiology and Public Health

Supplies:

- Calculators
- Access to the internet

Lab 12: Blood: the Good, the Bad, and the Ugly

Prepared slides and other supplies:

- Wrights-Giemsa stained blood smears (no pathology)
- Wrights-Giemsa or Giemsa stained blood smears of *Anaplasma* spp., *Babesia* spp., and *Dirofilaria* (microfilarial form)
- Blood cell counters (optional)
- Sure-Vue Mono Test kit (or other commercial test kit for Infectious Mononucleosis)

Appendix B

List of Potential Mythbusters Topics

The measles vaccine may cause vaccinated children to develop autism or other autism-related disorders like OCD and Asperger's syndrome.

Taking a probiotic pill containing bacterial cultures will improve your immune system and therefore your ability to fight off colds and other infections.

Bottled water is "cleaner" (meaning won't have any bacteria or viruses in it) and is therefore better for you to drink than regular tap water from a municipal source.

Getting an annual "flu shot" (influenza vaccine) means you will never have to worry about getting the flu.

If you drop something on the floor, it will still be safe to eat if you pick it up within 5 seconds of dropping it.

It is safe and economical to use a sponge in your kitchen for cleaning surfaces because they last for a long time.

Using a plastic cutting board in your kitchen to cut up food is safer than using a wooden cutting board.

Tuberculosis is a disease that we no longer have to worry about contracting here in the United States.

Children do not need to be immunized because most diseases preventable by vaccine (like mumps, measles, and polio) have been eliminated in the United States.

If you see any mold growing anywhere in your house, you should probably move out immediately because all molds are toxic.

Using hand soaps or household cleaning products labelled as "antibacterial" is healthier for you and your family.

Drinking lots of cranberry juice will prevent or cure urinary tract infections.

"Raw" or unpasteurized dairy products contain deadly microorganisms and should never be consumed by humans.

It's OK to let your dog "kiss" you on the mouth because a dog's mouth is cleaner than a human mouth.

Getting salad from a salad bar is perfectly safe because the plastic cover over the top protects the food from being contaminated.

You can catch genital herpes or chlamydia by sitting on a toilet seat in a public restroom.

Development of the symptoms of multiple sclerosis (MS) may be the result of an infection caused by bacteria or a virus.

Meat like ground beef and chicken should always be cooked "well done" because otherwise it is not safe to eat.

Using a product that contains zinc gluconate will shorten the duration of the common cold.

Using a product that contains the herb *Echinacea* will prevent you from getting a cold or the flu.

If you chew gum, your mouth will be "cleaner" (meaning, have fewer plaque-forming bacteria) than if you don't chew gum.

If you pick up a toad and it pees on your hand, you'll get a wart at the spot where the pee touches your skin.

Appendix C

Differentiation of Bacterial Cultures based on Morphological and Physiological Characteristics

Group 1: Aerobic and Anaerobic Cocci

Table 1. Aerobic cocci.		
	Gram stain	Arrangement
Micrococcus spp.	+	tetrads
Neisseria gonorrhoeae	–	diplococci

Table 2. Anaerobic cocci.				
	Catalase	Oxidase	Arrangement	Oxygen
Staphylococcus spp (see Table 1)	+	–	clusters	Facultative anaerobe
Streptococcus spp. (see Table 4)	–	–	chains	Aerotolerant anaerobe
Enterococcus spp. (see Table 4)	–	–	chains	Aerotolerant anaerobe

Table 3. Differentiation of *Staphylococcus* spp.			
	Staphylococcus aureus	*Staphylococcus epidermidis*	*Staphylococcus saprophyticus*
Coagulase	4+	–	1–2+
Mannitol Salt Agar	+	–	+
Urease	– or wk+	+	+
Hemolysis	Beta (β)	– or wk +	–
Novobiocin resistance	Susc. (-)	Susc. (-)	Res. (+)

Table 4. Differentiation of *Streptococcus* and *Enterococcus* spp.		
	Streptococcus pyogenes	*Enterococcus faecalis*
Hemolysis on BAP	Beta (β)	Gamma (γ)
Bile Esculin	No growth (-)	Growth/black (+)

Group 2: Aerobic and Facultatively Anaerobic Bacilli

Gram positive bacilli

Endospore forming: *Bacillus* spp. (see Table 5)

Non-endospore forming: *Corynebacterium xerosis*

Table 5. Differentiation of *Bacillus* spp.

	Endospore shape, location, sporangium
B. cereus	Oval, central, sporangium not swollen
B. lentus	Oval, terminal, sporangium swollen

Gram negative bacilli

Strictly aerobic: *Pseudomonas aeruginosa*

Facultatively anaerobic: Enterobacteriaceae (see Table 6)

Table 6. Differentiation of Enterobacteriaceae

	TSI	Sulfide/Indole/ Motility	Methyl Red	VP	Citrate
Escherichia coli	A/A	– /+/+	+	–	–
Enterobacter aerogenes	A/A	-/-/+	–	+	+
Citrobacter freundii	A/A	-/-/+	+	–	+
Klebsiella pneumoniae	A/A	-/-/-	–	+	+
Salmonella enteritidis	K/A	+/-/+	+	–	+

Image Credits

Lab 2: The Microscopic World

2-1: "Optical microscope nikon alphaphot +" by Moisey. Licensed under CC BY-SA 3.0 via Wikimedia Commons

2-2: "Condenser diagram" by Egmason. Licensed under CC BY-SA 3.0 via Wikimedia Commons

2-3: "Pencil in a bowl of water" by Theresa_knott, derivative work: Gregors. Licensed under CC BY-SA 3.0 via Wikimedia Commons

Lab 3: Bacteriological Culture Methods

3-1: Microbiological Media, photo by author

3-2: Streak Plate, photo by author

3-3: "Bacterial colony morphology". Licensed under Public Domain via Wikimedia Commons

3-4: "Pseudomonas aeruginosa and Staphylococcus aureus colonies" by HansN. Licensed under CC BY-SA 3.0 via Wikimedia Commons

3-5: Sterile pipettes, photo by author

3-6: Volume in pipette, photo by author

Lab 4: The Environmental Isolate Project

4-1: "Skin Microbiome20169-300" by Darryl Leja, NHGRI. Licensed under Public Domain via Wikimedia Commons

Lab 5: Differential Staining Techniques

5-1: "Cryptococcus neoformans using a light India ink staining preparation PHIL 3771 lores" by Photo Credit:Content Providers(s): CDC/Dr. Leanor Haley – This media comes from the Centers for Disease Control and Prevention's Public Health Image Library (PHIL), with identification number #3771. Licensed under Public Domain via Wikimedia Commons

5-2: "Staphylococcus aureus Gram" by Y Tambe. Licensed under CC BY-SA 3.0 via Wikimedia Commons

5-3, 5-4: "Bacterial morphology diagram" adapted by author from Mariana Ruiz LadyofHats. Licensed under Public Domain via Wikimedia Commons

5-5: "Gram stain 01" by Y tambe. Licensed under CC BY-SA 3.0 via Wikimedia Commons

5-6: "Mycobacterium tuberculosis Ziehl-Neelsen stain 640". Licensed under Public Domain via Wikimedia Commons

5-7: "Bacillus subtilis Spore" by Y tambe. Licensed under CC BY-SA 3.0 via Wikimedia Commons

5-8: "Paenibacillus alvei endospore microscope image" by TinaEnviro – Light microscope imaging. Licensed under CC BY-SA 3.0 via Wikimedia Commons –

5-8: "Bakterien Sporen" by Kookaburra. Licensed under Public Domain via Wikimedia Commons

Lab 6: Metabolism, Physiology, and Growth Characteristics of Cocci

6-1: "Catalase reaction" by Nase Licensed under CC BY-SA 3.0 via Wikimedia Commons

6-2: DrySlide Oxidase test, photo by author

6-3: Sequential chemical reactions, diagram by author

6-4: "TSIagar" by Y_tambe. Licensed under CC BY-SA 3.0 via Wikimedia Commons

6-5: Dichotomous key, diagram by author

6-6: MSA agar plate, photo by author

6-7: Beta hemolysis (β-hemolysis), photo by author

6-8: Gamma hemolysis (γ-hemolysis), photo by author

6-9: Urease test, photo by author

Lab 7: Metabolism, Physiology, and Growth Characteristics 0f Bacilli

7-1: SIM reactions, photo by author

Unnumbered: SIM interpretation, photo by author

7-2a: MR test, photo by author

7-2b: VP test, photo by author

7-3: Simmon's Citrate Agar plate, photo by author

Lab 9: Germ Warfare

9-1: "Zone of inhibition by microorganism bks" by Bijaysahu2013. Licensed under CC BY-SA 3.0 via Wikimedia Commons

Lab 11: Blood; the Good, the Bad, and the Ugly

11-1: "Hematopoesis EN" Michał Komorniczak. Licensed under CC BY 3.0 via Wikimedia Commons

11-2: "Hematopoiesis simple" by Mikael Häggström, from original by A. Rad. Licensed under CC BY-SA 3.0 via Wikimedia Commons

11-3: "Antibody" by Fvasconcellos. Licensed under Public Domain via Commons

11-4: "Peripheral blood smear – stained and unstained" by Coinmac. Licensed under CC BY-SA 3.0 via Wikimedia Commons

11-5: "Peripheral blood smear – stained and unstained" by Coinmac, cropped and annotated by author. Licensed under CC BY-SA 3.0 via Wikimedia Commons.

11-6: "Anaplasma phagocytophilum cultured in human promyelocytic cell line HL-60" by Kye-Hyung Kim, et al., EID Journal,Volume 20, Number 10, http://wwwnc.cdc.gov/eid/article/20/10/13-1680-f2. Licensed under Public Domain via Wikimedia Commons

11-7: "Babesia life cycle human en" by LadyofHats Mariana Ruiz Villarreal. Licensed under Public Domain via Wikimedia Commons

11-8: "Babiesa spp" by CDC/ Steven Glenn; Laboratory & Consultation Division – This media comes from the Centers for Disease Control and Prevention's Public Health Image Library (PHIL), with identification number #5943. Licensed under Public Domain via Wikimedia Commons

11-9: "Dirofilaria immitis lifecycle" by Anka Friedrich, Cú Faoil (text). Licensed under CC BY-SA 3.0 via Wikimedia Commons

11-10: "Microfilaria" by Joelmills. Licensed under CC BY-SA 3.0 via Wikimedia Commons